OFFICIAL SQA PAST PAPERS WITH ANSWERS

HIGHER

# GEOGRAPHY
## 2006-2009

BrightRED
PUBLISHING

© Scottish Qualifications Authority

First exam published in 2006.
Published by Bright Red Publishing Ltd, 6 Stafford Street, Edinburgh EH3 7AU
tel: 0131 220 5804 fax: 0131 220 6710 info@brightredpublishing.co.uk  www.brightredpublishing.co.uk

ISBN 978-1-84948-059-8

A CIP Catalogue record for this book is available from the British Library.

Bright Red Publishing is grateful to the copyright holders, as credited on the final page of the book, for permission to use their material.
Every effort has been made to trace the copyright holders and to obtain their permission for the use of copyright material.
Bright Red Publishing will be happy to receive information allowing us to rectify any error or omission in future editions.

# 2006

[BLANK PAGE]

# X208/301

NATIONAL
QUALIFICATIONS
2006

MONDAY, 29 MAY
9.00 AM – 10.30 AM

GEOGRAPHY
HIGHER
Paper 1
Physical and
Human Environments

**Six** questions should be attempted, namely:

**all four** questions in **Section A** (Questions 1, 2, 3 and 4);

**one** question from **Section B** (Question 5 **or** Question 6);

**one** question from **Section C** (Question 7 **or** Question 8).

Write the numbers of the **six** questions you have attempted in the marks grid on the back cover of your answer booklet.

The value attached to each question is shown in the margin.

Credit will be given for appropriate maps and diagrams, and for reference to named examples.

Questions should be answered in sentences.

**Note** The reference maps and diagrams in this paper have been printed in black only: no other colours have been used.

SCOTTISH
QUALIFICATIONS
AUTHORITY

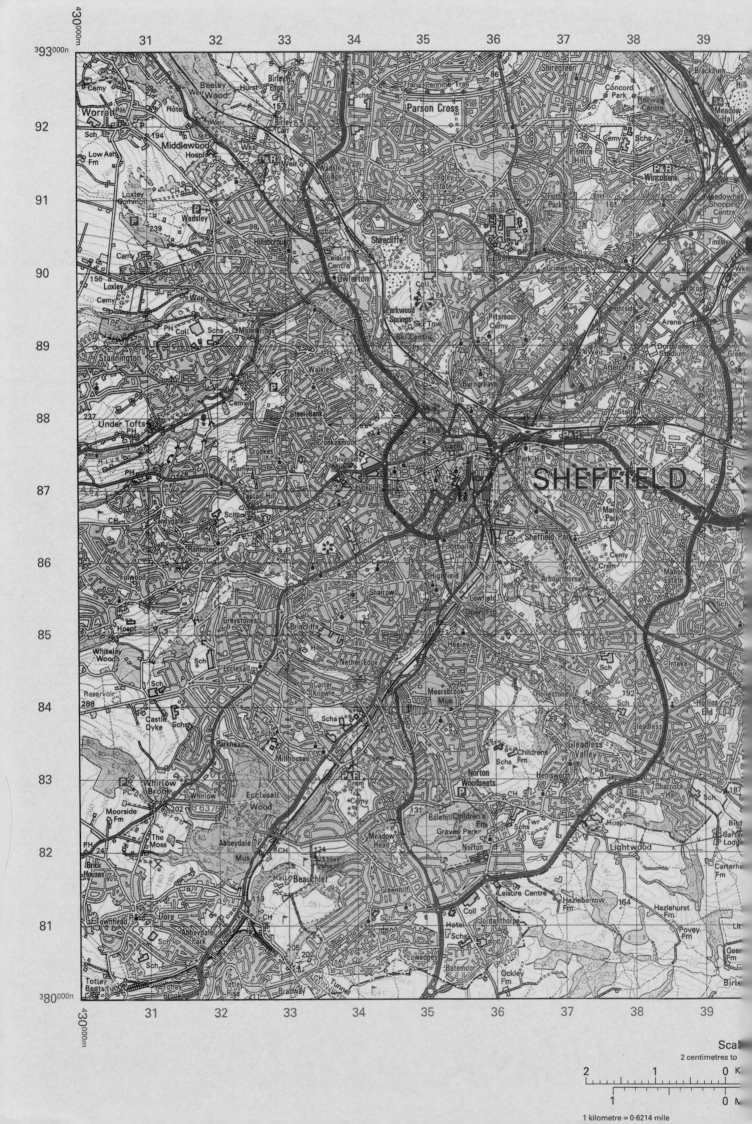

*Marks*

**SECTION A: Answer ALL questions in this section**

**Question 1: Atmosphere**

(a) Study Reference Diagram Q1.

**Explain** why the Earth's surface can absorb only 50% of the solar energy received at the outer atmosphere.

**4**

(b) With the aid of an annotated diagram, **explain** why there is a surplus of solar energy in the tropical latitudes and a deficit of solar energy towards the poles.

**5**

**Reference Diagram Q1 (Earth—atmosphere energy exchanges)**

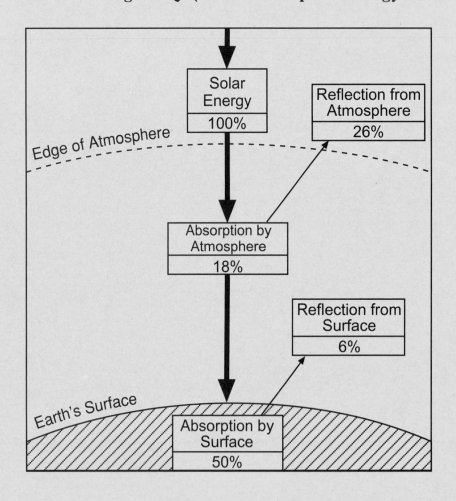

*Marks*

## Question 2: Lithosphere

(*a*)  Study Reference Diagram Q2.

The landscape shown in Reference Diagram Q2 contains features of coastal **erosion**.

**Explain** the formation of this coastal landscape, referring in detail to the processes involved.  Annotated diagrams may be used.

**6**

**Reference Diagram Q2 (Coastal erosion at The Foreland, Dorset)**

**[Turn over**

*Marks*

**Question 2 (continued)**

(*b*)  Study Reference Photograph Q2.

"*Many coastal slopes are subject to mass movements.*"

**Describe** the conditions **and** processes that have led to the slumping shown in
the photograph.

3

**Reference Photograph Q2 (Holderness, Yorkshire)**

*Marks*

## Question 3: Rural Geography

Study Reference Photograph Q3.

(a) **Describe** and **explain** the main features of a commercial arable farming landscape.

3

(b) "*There have been many changes to commercial arable farming over the last 50 years. Whilst many have benefited farmers, others have been less welcome and some have caused problems for the environment.*"

Referring to a **named area** of commercial arable farming which you have studied:

(i)  **describe** the changes which have taken place;

(ii) **discuss** the impact of such changes on the people **and** the local environment.

5

**Reference Photograph Q3 (Commercial arable farming landscape)**

**[Turn over**

*Marks*

## Question 4:  Urban Geography

Study OS Map Extract number 1491/111: Sheffield (*separate item*), and Reference Map Q4.

(*a*) What map evidence suggests that the Central Business District of Sheffield mostly lies in squares 3586 and 3587?    **3**

(*b*) Meadowhall Shopping Centre (GR 3990, 3991)—shown on Reference Map Q4— is one of the largest regional indoor shopping centres in the UK.

With the aid of map evidence, **describe** and **explain** the advantages of its location.    **4**

(*c*) **Suggest** the impact which Meadowhall Shopping Centre may have had on the traditional Central Business District of Sheffield.    **3**

### Reference Map Q4

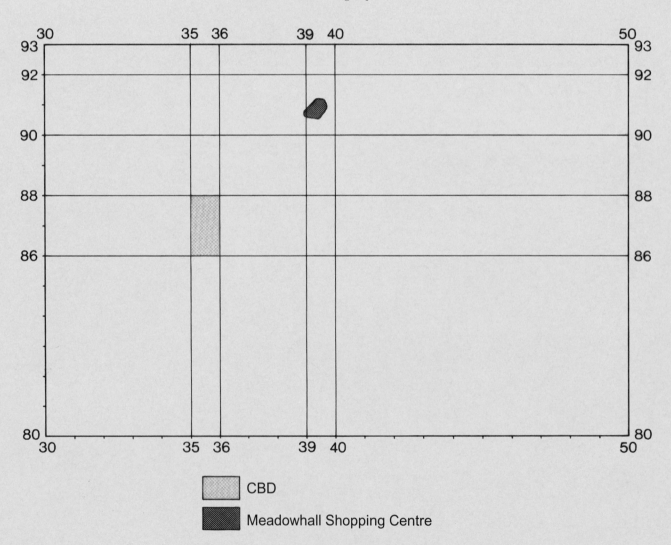

*Marks*

### SECTION B:  Answer ONE question from this section, ie either Question 5 or Question 6.

## Question 5:  Hydrosphere

(*a*) "*A drainage basin is an open system with four elements—**inputs, storage, transfers** and **outputs**.*"

**Describe** the movement of water within a drainage basin.    **4**

(*b*) Study Reference Diagrams Q5A and Q5B.

**Describe** and **explain** the changing river levels at Stratford-on-Avon between 5 and 13 April 1998.    **3**

### Reference Diagram Q5A (Rainfall data for Stratford-on-Avon)

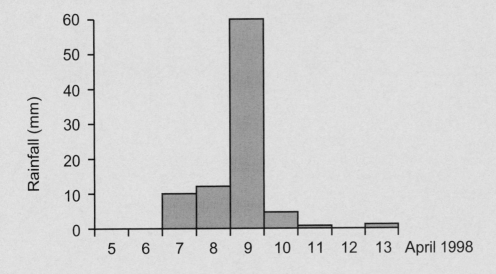

### Reference Diagram Q5B (Changing river levels at Stratford-on-Avon)

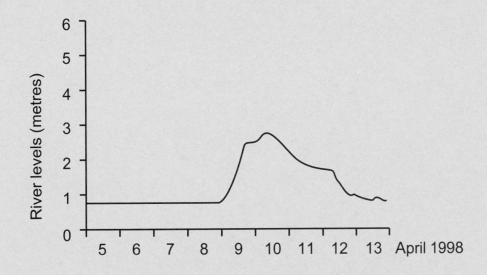

*Marks*

**DO NOT ANSWER THIS QUESTION IF YOU HAVE
ALREADY ANSWERED QUESTION 5**

### Question 6:  Biosphere

Study Reference Diagram Q6 which shows soil profiles for a podzol and a brown earth.

(a) **Describe** the different properties (horizons, colour, texture, drainage) of the two soils shown.

(b) **Explain** the differences in their formation.                                    7

**Reference Diagram Q6 (Selected soil profiles)**

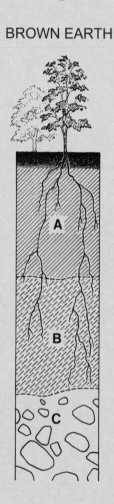

PODZOL                        BROWN EARTH

*Marks*

SECTION C:  Answer ONE question from this section,
ie either Question 7 or Question 8.

### Question 7:  Population Geography

Study Reference Diagrams Q7A and Q7B which show predicted changes in Scotland's future population.

(*a*) **Suggest reasons** for the changes shown.    4

(*b*) Scottish politicians have suggested that one of the implications of these changes will be a need for more migrants to be attracted to Scotland.

Referring to Scotland, **or** any other EMDC (Economically More Developed Country) which you have studied, **discuss** the likely **benefits** that this could bring.    3

**Reference Diagram Q7A (Predicted change in Scotland's population 2002–2032)**

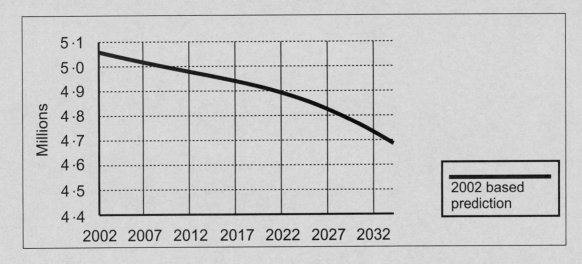

**Reference Diagram Q7B (Predicted change in Scotland's population aged 65+, 2002–2032)**

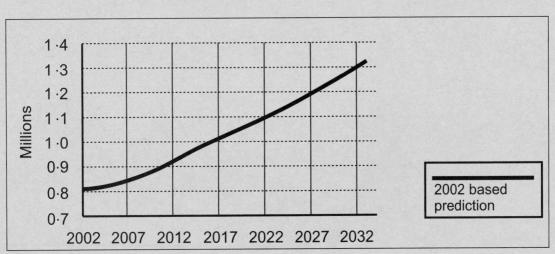

*Marks*

**DO NOT ANSWER THIS QUESTION IF YOU HAVE
ALREADY ANSWERED QUESTION 7**

**Question 8:  Industrial Geography**

Study OS Map Extract number 1491/111: Sheffield (*separate item*), and Reference Map Q8.

(*a*) Area A on Reference Map Q8 was once the centre of Sheffield's traditional metal-working industry.

With the aid of map evidence, **describe** the likely **old** industrial landscape of this area.

3

**Reference Map Q8**

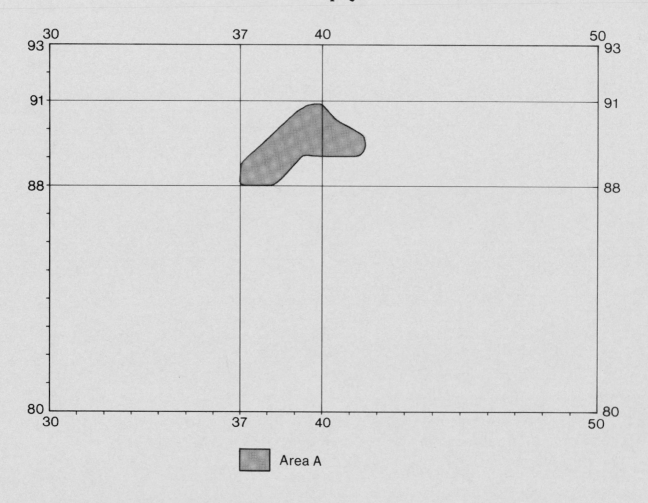

Area A

*Marks*

**Question 8 (continued)**

(b)  Study Reference Table Q8.

*"The changes in employment figures for Sheffield are typical of those of many old, traditional, industrial areas."*

For **either** Sheffield **or** any other industrial concentration in the European Union which you have studied, **explain** why such changes in employment have taken place.                                                                      **4**

**Reference Table Q8   (Major employment changes in Sheffield 1971–2001)**

| Number of jobs in: | 1971 | 2001 | % Change in 30 years |
|---|---|---|---|
| All types of manufacturing | 136 000 | 36 000 | −73·5% |
| Metal products and basic metal manufacturing | 90 000 | 17 000 | −81% |
| Services | 140 000 | 194 000 | +38% |

*[END OF QUESTION PAPER]*

[BLANK PAGE]

# X208/303

NATIONAL
QUALIFICATIONS
2006

MONDAY, 29 MAY
10.50 AM – 12.05 PM

GEOGRAPHY
HIGHER
Paper 2
Environmental
Interactions

**Two** questions should be attempted, namely:

**one** question from **Section 1** (Questions 1, 2, 3) and
**one** question from **Section 2** (Questions 4, 5, 6).

Write the numbers of the **two** questions you have attempted in the marks grid on the back cover of your answer booklet.

The value attached to each question is shown in the margin.

Credit will be given for appropriate maps and diagrams, and for reference to named examples.

Questions should be answered in sentences.

**Note**   The reference maps and diagrams in this paper have been printed in black only:  no other colours have been used.

SCOTTISH
QUALIFICATIONS
AUTHORITY

*Marks*

## SECTION 1

**You must answer ONE question from this Section.**

**Question 1** (Rural Land Resources)

(a) *"Snowdonia is an area of outstanding glaciated upland scenery that attracts up to 10 million visitors every year."*

   **Describe** and **explain**, with the aid of annotated diagrams, the formation of the glacial features in Snowdonia **or** in any other glaciated upland area which you have studied.    **10**

(b) With reference to Snowdonia **or** any upland area you have studied, **explain** the social and economic opportunities created by the landscape.    **5**

(c) 

> *"The Welsh Highland Railway route runs through some of the most stunning scenery in Britain . . . the line climbs into the heart of the Snowdonia mountains, past lakes, then down through a forest and a spectacular rocky gorge to the sea at Porthmadog . . . The rebuilt line will also offer a new way to reach the communities and countryside en route, and offer an alternative to motor transport in an ecologically sensitive area, particularly for the tens of thousands who visit Snowdonia every summer."*

   This is a quote from the "Welsh Highland Railway" website. However, the rebuilding of the Welsh Highland Railway has caused controversy.

   Study Reference Map Q1. **Explain** how the construction and operation of the Welsh Highland Railway might be favoured by some groups and opposed by others.    **6**

(d) For Snowdonia, **or** any other upland area you have studied, **describe** the measures taken to resolve environmental conflicts related to tourism.    **4**

   **(25)**

*Marks*

## Question 1 – continued

### Reference Map Q1 (The route of the Welsh Highland Railway)

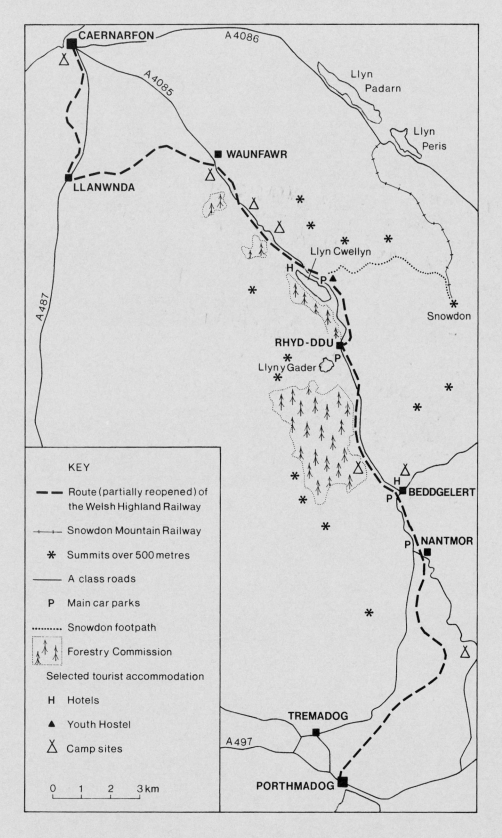

KEY

- **– – –** Route (partially reopened) of the Welsh Highland Railway
- **++++** Snowdon Mountain Railway
- **\*** Summits over 500 metres
- **——** A class roads
- **P** Main car parks
- **······** Snowdon footpath
- **↟↟↟** Forestry Commission

Selected tourist accommodation

- **H** Hotels
- **▲** Youth Hostel
- **△** Camp sites

0  1  2  3 km

**[Turn over**

*Marks*

**Question 2** (Rural Land Degradation)

Study Reference Diagram Q2.

(a)  (i)  **Describe** the rainfall pattern illustrated by the graph.     **3**

(ii)  **Explain** why such a rainfall pattern could lead to the degradation of rural land.     **4**

(b)  **Describe** the impact of rural land degradation on the people and environment in named areas of North America **and either** Africa north of the Equator **or** the Amazon Basin.     **8**

**Reference Diagram Q2 (Rainfall variability in the Sahel)**

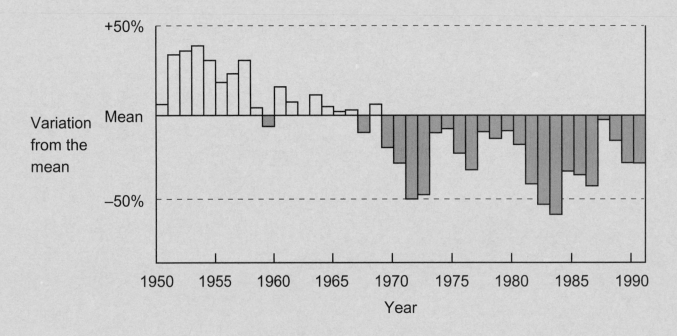

*Marks*

## Question 2 – continued

(*c*)  Study Reference Table Q2.

Select **two** soil conservation strategies from North America **and two** from **either** Africa north of the Equator **or** the Amazon Basin.

**Describe** your chosen methods and **explain** how they help to conserve soil in rural areas.

**10**

**(25)**

### Reference Table Q2 (Soil conservation strategies)

| | North America | | Africa north of the Equator | | The Amazon Basin |
|---|---|---|---|---|---|
| 1 | Shelter belts | 1 | "Magic Stones" (Diguettes) | 1 | Return land to traditional farming |
| 2 | Strip cropping | 2 | Dams built in gullies | 2 | Purchase by conservation groups |
| 3 | Contour ploughing | 3 | Animal fences | 3 | Crop rotation |
| 4 | Land reclamation | 4 | Dune stabilisation | 4 | Agro-forestry schemes |

**[Turn over**

*Marks*

**Question 3** (River Basin Management)

Study Reference Maps Q3A and Q3B, Reference Diagrams Q3A and Q3B and Reference Table Q3 which show information about the Three Gorges Dam Project in China.

(*a*) **Explain** why there is a need for water management on the Yangtze River.    **6**

(*b*) With reference to the Three Gorges Dam Project **or** any water management project you have studied in North America **or** Africa **or** Asia, **explain** the human and physical factors that have to be considered when selecting sites for dams and their associated reservoirs.    **7**

(*c*) For your chosen water management project, **describe** and **explain** the social, economic and environmental benefits and adverse consequences of the project.    **12**

    **(25)**

**Reference Map Q3A (The Yangtze Basin and location of the Three Gorges Dam)**

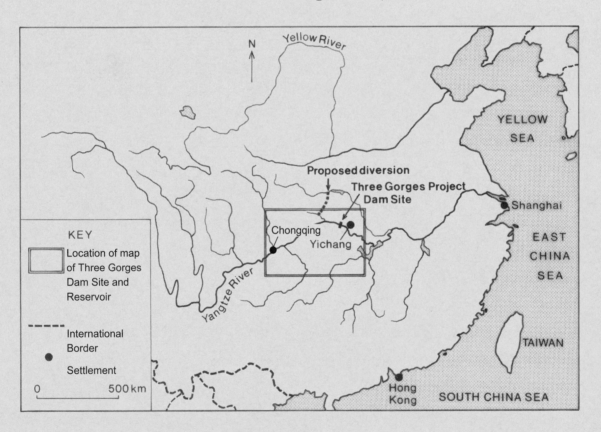

**Question 3 – continued**

**Reference Map Q3B (The Three Gorges Dam site and reservoir)**

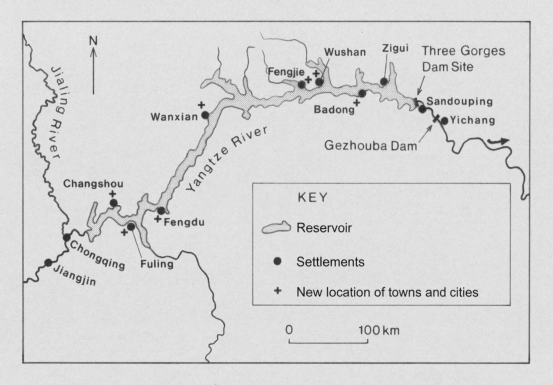

**Reference Diagram Q3A (Yichang monthly rainfall graph)**

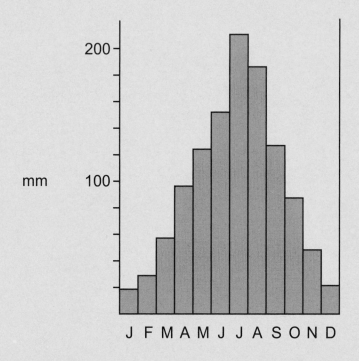

**[Turn over**

**Question 3 – continued**

### Reference Diagram Q3B (Average monthly river flow at Yichang)

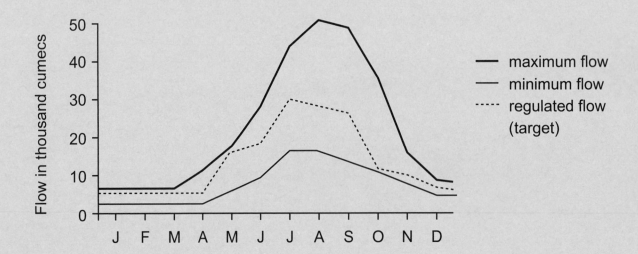

### Reference Table Q3 (Population growth in China—actual and projected)

| Year | Total (millions) |
|------|------------------|
| 1950 | 557 |
| 1995 | 1227 |
| 2010 | 1380 |
| 2025 | 1488 |

**[Turn over for Question 4 on *Page ten***

*Marks*

## SECTION 2

**You must answer ONE question from this Section.**

**Question 4** (Urban Change and its Management)

Study Reference Map Q4.

(*a*) **Describe** and **account for** the distribution of major cities in **either** Germany **or** any other EMDC (Economically More Developed Country) that you have studied.　　**4**

(*b*) For any city you have studied in an EMDC, **explain** the ways in which its site and situation contributed to its growth.　　**4**

(*c*) In many cities in EMDCs people have been moving from areas of high density housing in the inner city to areas of low density housing in the suburbs or commuter villages.

For a city that you have studied in an EMDC:

(i) **explain** the reasons for this outward movement of population;

(ii) **outline** ways in which this city has tried to deal with the problems caused by this population movement.　　**7**

(*d*) Many cities in ELDCs (Economically Less Developed Countries) continue to grow in both size and population.

**Explain** the reasons behind this growth.　　**5**

(*e*) Shanty towns are one result of this rapid growth of population. **Outline** the social and environmental **problems** of shanty towns and highlight any **advantages** they might have for their inhabitants.　　**5**

**(25)**

## Question 4 - continued

**Reference Map Q4 (German cities with over 500 000 inhabitants)**

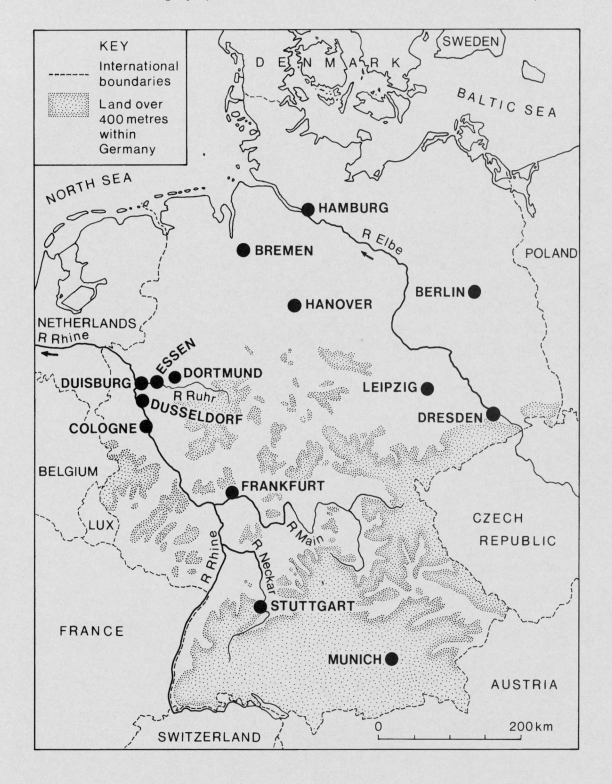

[Turn over

*Marks*

**Question 5** (European Regional Inequalities)

(*a*)  Study Reference Map Q5.

The shaded area on the map has been referred to as the "Euro-Core".

**Suggest** both physical and human reasons for the prosperity enjoyed in this part of Europe.

6

(*b*)  Study Reference Table Q5 which shows a range of indicators of development for five of the ten countries which joined the European Union on 1 May 2004.

In what ways does the data provide evidence of these newest member countries being less wealthy than the group of 15 countries which made up the EU before that date?

5

(*c*)  "*The enlarged European Union continues to have large variations in levels of wealth and economic development within individual countries.*"

For a named country in the European Union which has marked differences in economic development **within** it:

(i)   **describe** some of the social and economic problems which can arise from such regional differences;

6

(ii)  **discuss** ways in which less prosperous regions can receive help from the EU **and** their national government to overcome such problems.

8

**(25)**

**Question 5 – continued**

## Reference Map Q5 (The "Euro-Core")

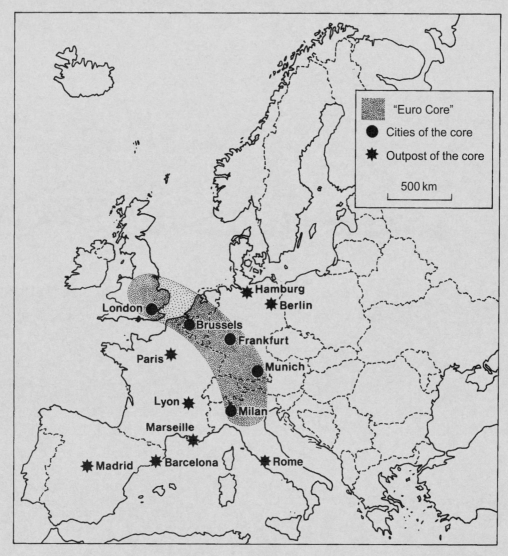

### Reference Table Q5 (Selected indicators of development for 5 of the 10 countries which joined the EU on 1 May 2004)

|                 | A     | B   | C   | D   |
|-----------------|-------|-----|-----|-----|
| **EU 15 average** | 8·1 | 31  | 85  | 100 |
| Cyprus          | 4·3   | 22  | 55  | 78  |
| Czech Republic  | 7·5   | 14  | 84  | 61  |
| Hungary         | 5·8   | 10  | 68  | 51  |
| Lithuania       | 13·4  | 7   | 47  | 37  |
| Poland          | 19·3  | 9   | 36  | 41  |

**A**   Unemployment rate (%) (2003)

**B**   Number of personal computers per 100 people (2001)

**C**   Number of mobile phone subscribers per 100 people (2002)

**D**   Gross Domestic Product (GDP) per capita as a % of EU 15 Average

(EU 15 = the 15 existing member countries before 1 May 2004)

*Marks*

**Question 6** (Development and Health)

Study Reference Table Q6A.

(*a*) Reference Table Q6A shows data for three countries which, although ELDCs (Economically Less Developed Countries), are relatively rich.

Giving examples from named countries, **describe** the factors which help some ELDCs to achieve higher levels of development than others.    **5**

(*b*) Study Reference Table Q6B and Reference Map Q6.

  (i) **Suggest** the physical and human factors which might lead to ill-health and early death among many of the people of Mali.    **5**

  (ii) Primary Health Care is one way of improving health in countries such as Mali.  For Mali **or** any other ELDC you have studied, **describe** the main strategies involved in Primary Health Care.    **6**

(*c*) For malaria **or** bilharzia **or** cholera, **describe** the measures which can be taken to combat the disease and **assess** their effectiveness.    **9**

    **(25)**

**Reference Table Q6A (Development indicators for selected countries)**

| Country | Birth Rate (per 1000) | Infant Mortality Rate (per 1000 live births) | Literacy Rate (%) | GNP per capita ($US) |
|---|---|---|---|---|
| Thailand | 16 | 21 | 93 | 7400 |
| Saudi Arabia | 30 | 14 | 79 | 11 800 |
| South Korea | 12 | 7 | 98 | 17 700 |

**Question 6 – continued**

**Reference Table Q6B (Mali—selected information)**

| | |
|---|---|
| % of land surface as arable (crop) land | 4 |
| Birth rate per 1000 | 47 |
| Literacy rate (%) | 46 |
| % employed in farming and fishing | 80 |
| GDP per capita ($US) | 900 |
| % population below the poverty line | 64 |

**Reference Map Q6 (Mali)**

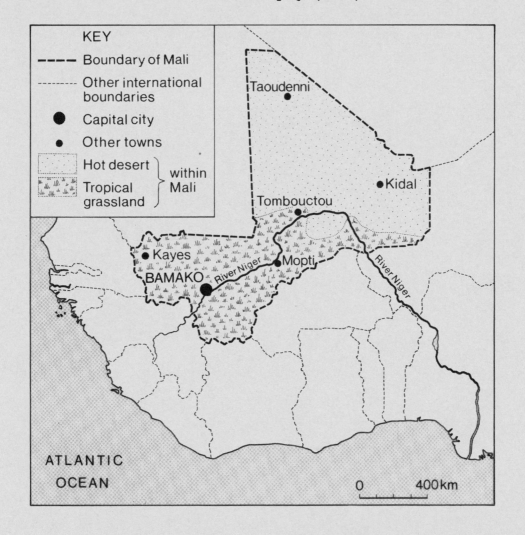

[END OF QUESTION PAPER]

[BLANK PAGE]

[BLANK PAGE]

# X208/301

NATIONAL
QUALIFICATIONS
2007

MONDAY, 28 MAY
9.00 AM – 10.30 AM

GEOGRAPHY
HIGHER
Paper 1
Physical and
Human Environments

**Six** questions should be attempted, namely:

**all four** questions in **Section A** (Questions 1, 2, 3 and 4);

**one** question from **Section B** (Question 5 **or** Question 6);

**one** question from **Section C** (Question 7 **or** Question 8).

Write the numbers of the **six** questions you have attempted in the marks grid on the back cover of your answer booklet.

The value attached to each question is shown in the margin.

Credit will be given for appropriate maps and diagrams, and for reference to named examples.

Questions should be answered in sentences.

**Note**    The reference maps and diagrams in this paper have been printed in black only:  no other colours have been used.

SCOTTISH
QUALIFICATIONS
AUTHORITY

©

1:50 000 Scale
Landranger Series

Four colours should appear
Four colours should appear

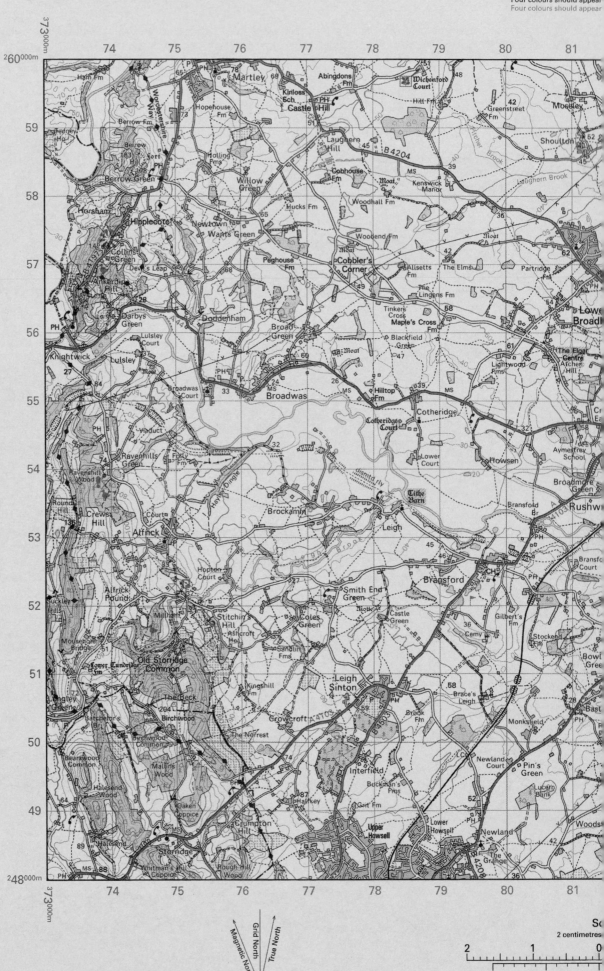

Diagrammatic only

Magnetic North   Grid North   True North

So

2 centimetres

1 kilometre = 0·6214 mile

please return to the invigilator.
e grid return to the invigilator.

1 mile = 1·6093 kilometres

*Marks*

### SECTION A:  Answer ALL questions in this section

## Question 1:  Atmosphere

(*a*)  Study Reference Diagram Q1A and Reference Map Q1A.

Identify air masses A and B, and **describe** their origin and nature.     3

**Reference Diagram Q1A (The Inter-tropical Convergence Zone (ITCZ))**

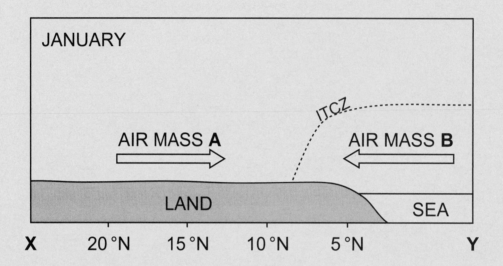

**Reference Map Q1A (Location of section X–Y)**

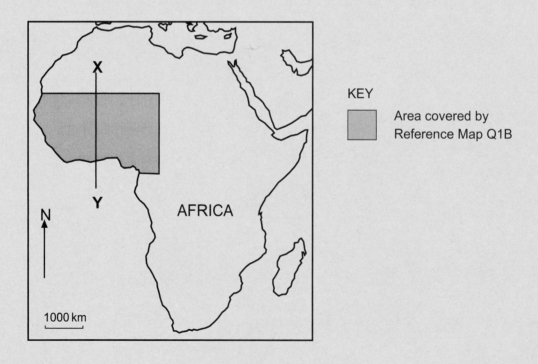

*Marks*

## Question 1 (continued)

(b)  Study Reference Map Q1B and Reference Diagram Q1B.

**Describe** and **explain** the varying rainfall patterns shown in Reference
Diagram Q1B.

6

### Reference Map Q1B (Rainfall patterns in West Africa)

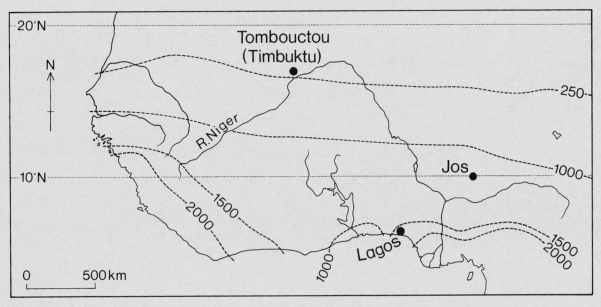

KEY  ------------ Isohyets showing mean annual rainfall (mm)

~~~~~ Rivers

### Reference Diagram Q1B (West Africa—selected rainfall graphs)

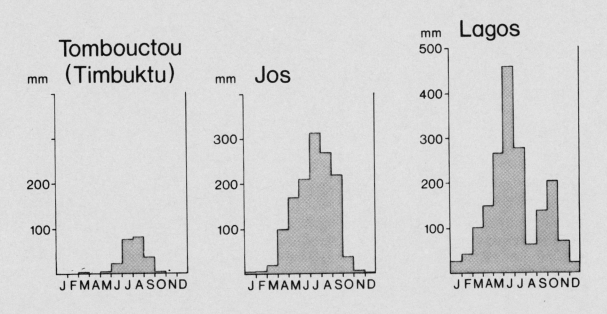

[Turn over

*Marks*

**Question 2:  Hydrosphere**

Study OS Map Extract number 1560/150: Worcester (*separate item*).

(*a*)  The River Teme is in the lower section of its course between 760545 and
850522.  **Describe** the physical characteristics of the river and its valley in this
section.                                                                                5

(*b*)  Select **one** of the features that you have described for part (*a*) and **explain**,
with the aid of a diagram or diagrams, how this feature is formed.                       4

*Marks*

## Question 3:  Population Geography

(*a*)  **Describe** the ways in which countries can obtain accurate population data.          **3**

(*b*)  **Explain**

    (i)   why **ELDCs** (Economically Less Developed Countries) may find the collection of such data more difficult, and

    (ii)  why the quality of data obtained may be less reliable than that gathered in an **EMDC** (Economically More Developed Country).          **6**

**[Turn over**

*Marks*

## Question 4:  Industrial Geography

(*a*)  For South Wales, or any other industrial concentration in the European Union which you have studied, **describe** the physical and human factors which led to the growth of traditional industries before 1950.

**4**

(*b*)  Study Reference Map Q4.

**Describe** and **explain** the methods used to attract newer industries and investments to South Wales, or to any other industrial concentration in the European Union which you have studied.

**5**

**Reference Map Q4 (South Wales:  Location of industrial developments since 1950)**

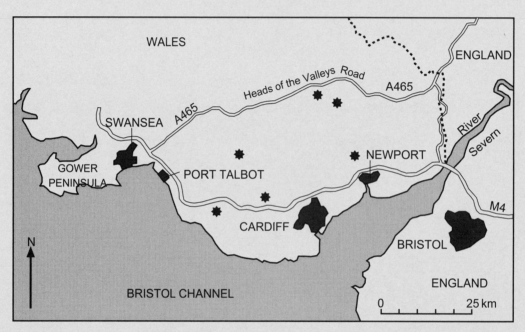

* Location of industrial developments since 1950

*Marks*

**SECTION B: Answer ONE question from this section,
ie either Question 5 or Question 6.**

**Question 5: Lithosphere**

Study Reference Photograph Q5 which shows a glaciated upland landscape in the Cairngorm Mountains.

(*a*) **Describe** the evidence which suggests that the area shown in the photograph has been affected by the processes of glacial erosion.    3

(*b*) Choose **one** feature of glacial erosion visible in the photograph and, with the aid of an annotated diagram (or diagrams), **explain** how it was formed.    4

**Reference Photograph Q5**

**[Turn over**

*Marks*

**DO NOT ANSWER THIS QUESTION IF YOU HAVE
ALREADY ANSWERED QUESTION 5**

## Question 6: Biosphere

Study Reference Diagram Q6A which shows two soil profiles.

Choose **one** of the soil profiles.

(i) **Describe** the characteristics of the soil, including horizons, colour, texture and
drainage.

3

**Reference Diagram Q6A (Selected soil profiles)**

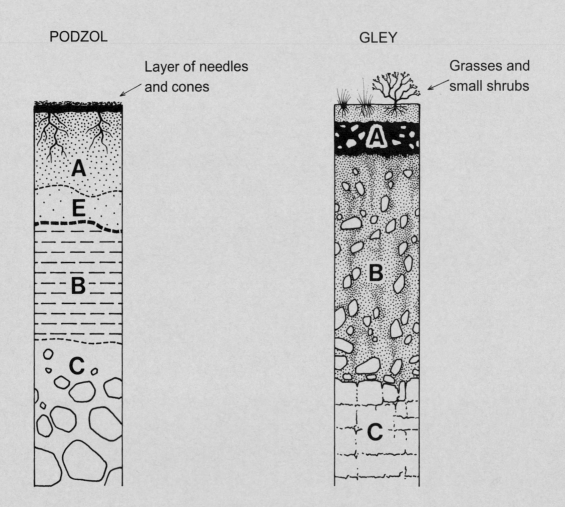

*Marks*

**Question 6 (continued)**

(ii) Study Reference Diagram Q6B.

**Explain** how the major soil forming factors shown in the diagram have contributed to the formation of your chosen soil profile.

4

**Reference Diagram Q6B (Main factors affecting soil formation)**

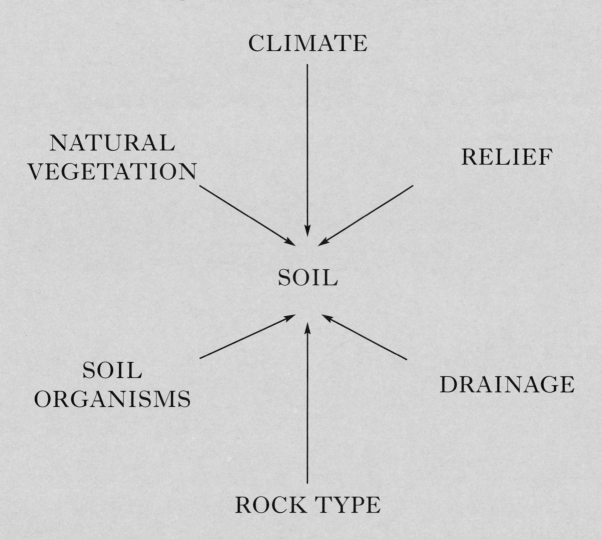

**[Turn over**

*Marks*

**SECTION C: Answer ONE question from this section,**
**ie either Question 7 or Question 8.**

## Question 7: Rural Geography

(a) **Describe** the main characteristics of **shifting cultivation**.

3

(b) *"In Central America, population density and loss of rainforest cover are closely related. During the last two decades, human activities have caused the deforestation of more than 120 000 square kilometres each year."*

Referring to a named area where shifting cultivation is carried out, **explain** the **impact** which deforestation and increased population density have had on the environment and way of life of the shifting cultivators.

4

**DO NOT ANSWER THIS QUESTION IF YOU HAVE**
**ALREADY ANSWERED QUESTION 7**

## Question 8: Urban Geography

Study OS Map Extract number 1560/150: Worcester (*separate item*), and Reference Map Q8.

(a) **Describe** the urban environment of Area A and **explain** its location.

4

(b) For **either** Area B **or** Area C, **explain** the advantages of the residential environment.

3

**Question 8 (continued)**

**Reference Map Q8 (Location of urban areas in Worcester)**

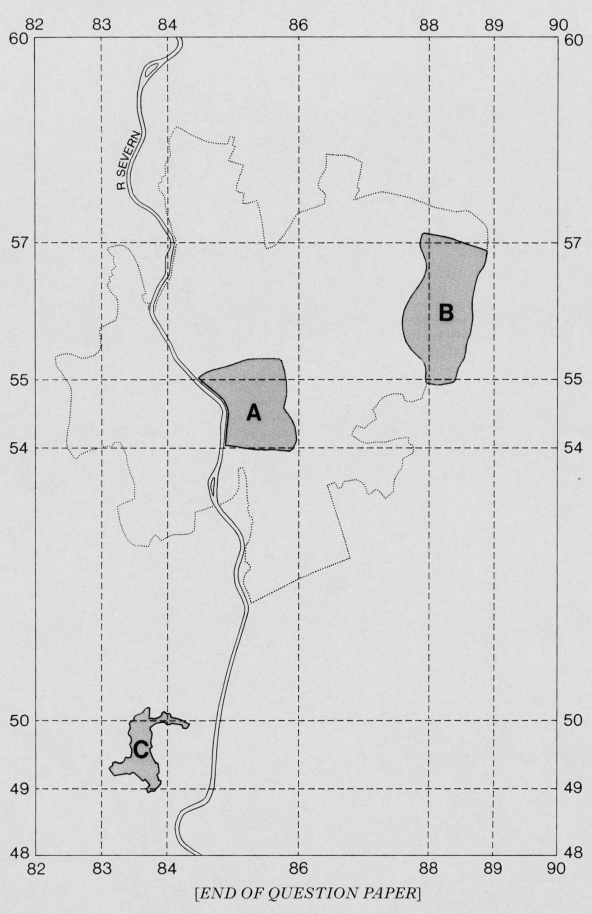

[END OF QUESTION PAPER]

[BLANK PAGE]

# X208/303

NATIONAL
QUALIFICATIONS
2007

MONDAY, 28 MAY
10.50 AM – 12.05 PM

GEOGRAPHY
HIGHER
Paper 2
Environmental
Interactions

**Two** questions should be attempted, namely:

**one** question from **Section 1** (Questions 1, 2, 3) and
**one** question from **Section 2** (Questions 4, 5, 6).

Write the numbers of the **two** questions you have attempted in the marks grid on the back cover of your answer booklet.

The value attached to each question is shown in the margin.

Credit will be given for appropriate maps and diagrams, and for reference to named examples.

Questions should be answered in sentences.

**Note**    The reference maps and diagrams in this paper have been printed in black only: no other colours have been used.

SCOTTISH
QUALIFICATIONS
AUTHORITY

*Marks*

## SECTION 1

**You must answer ONE question from this Section.**

**Question 1** (Rural Land Resources)

(*a*)  Study Reference Maps Q1A and Q1B.

**Describe** and **suggest reasons for** the location of Britain's National Parks.    5

(*b*)  Study Reference Diagram Q1 on *Page four*.

*"Tourism can bring benefits but also causes problems for National Parks."*

With the aid of Reference Diagram Q1 and referring to a specific National Park **or** other named upland **or** coastal landscape area which you have studied:

(i)  **describe** some of the benefits which an influx of tourists has brought; and

(ii)  **suggest** and **evaluate** ways in which the **negative** effects of tourism can be tackled.    10

(*c*)  With the aid of annotated diagrams, **describe** and **explain** the formation of the main features of any **coastal landscapes** which you have studied. You should refer to erosional **and** depositional features in your answer.    10

    **(25)**

**Question 1 – continued**

**Reference Map Q1A
(National Parks in Great Britain)**  **Reference Map Q1B
(Relief map of Great Britain)**

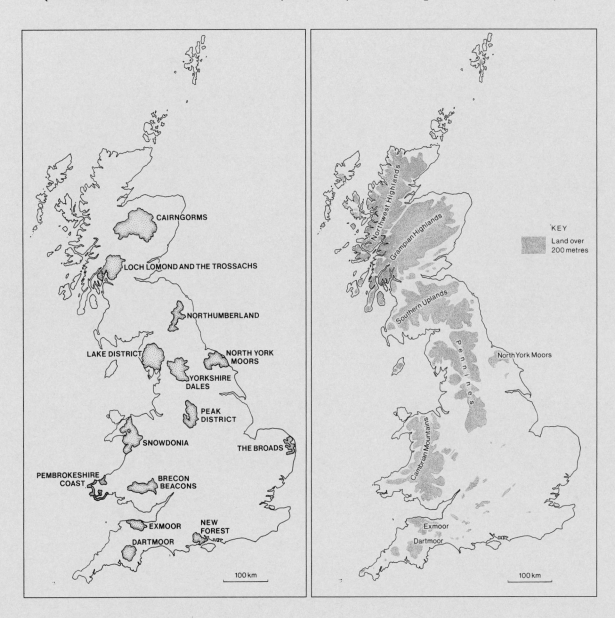

[Turn over

**Question 1 – continued**

**Reference Diagram Q1 (Positive and negative effects of tourism)**

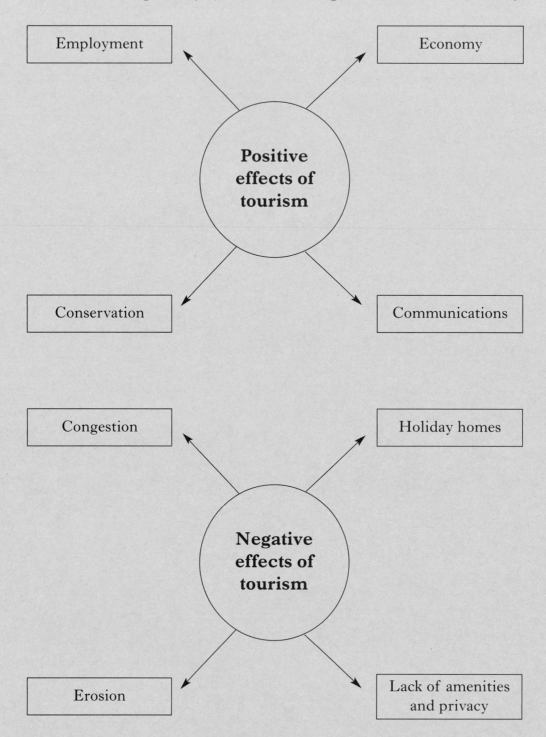

*Marks*

**Question 2** (Rural Land Degradation)

(*a*)  Study Reference Table Q2.

Describe the processes of soil erosion by both water and wind.    **5**

(*b*)  Describe and explain the main human causes of land degradation in North America **and either** Africa north of the Equator **or** the Amazon Basin.    **8**

(*c*)  Referring to named locations in **either** Africa north of the Equator **or** the Amazon Basin, describe the social and economic impact of land degradation on the people.    **5**

(*d*)  Study Reference Diagram Q2.

For any **four** methods of soil conservation, explain how each helps to conserve soil and reduce land degradation.    **7**

**(25)**

**Reference Table Q2 (Types of water and wind erosion)**

| Soil Erosion by Water | Soil Erosion by Wind |
|---|---|
| Rainsplash | Suspension |
| Sheet wash | Saltation |
| Rill erosion | Surface creep |
| Gully erosion | |

**Reference Diagram Q2 (Soil conservation strategies)**

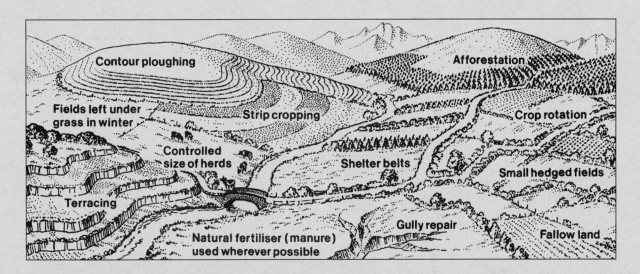

**[Turn over**

*Marks*

**Question 3** (River Basin Management)

(a) Study Reference Maps Q3A, Q3B and Q3C.

For Asia **or** Africa **or** North America, **describe** and **explain** the general distribution of the main river basins.

5

(b) "*Among the 30 largest dams planned for the Narmada river in India, the Sardar Sarovar is the largest. Most of the water held back by the dam will be used in the neighbouring state of Gujarat as it expands its rice and cotton production.*"

Study Reference Maps Q3D, Q3E and Reference Diagram Q3 on *Page eight*.

**Explain** why there is a need for water management in the Narmada River Basin and in Gujarat State.

5

(c) For the Narmada River Project **or** any other river basin management project in Asia **or** Africa **or** North America, **explain** the political problems that may have resulted from the project.

3

(d) **Describe** and **suggest reasons for** the social, economic and environmental benefits **and** adverse consequences of a named water control project in Asia **or** Africa **or** North America.

12

(25)

## Question 3 – continued

### Reference Map Q3A
### (Major river basins of Africa)

### Reference Map Q3B
### (Major river basins of North America)

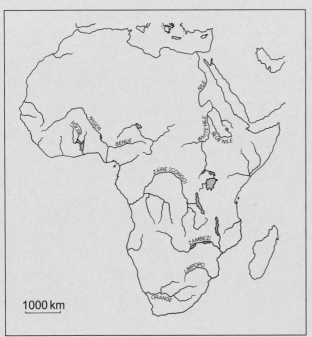

### Reference Map Q3C (Major river basins of Asia)

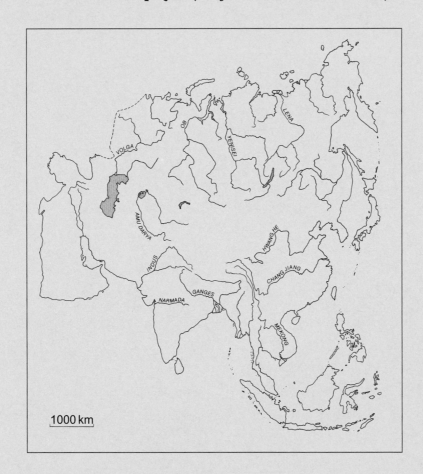

## Question 3 – continued

**Reference Map Q3D**
**(Location of Narmada River Basin in India)**

**Reference Diagram Q3**
**(Climate graphs for Ahmedabad)**

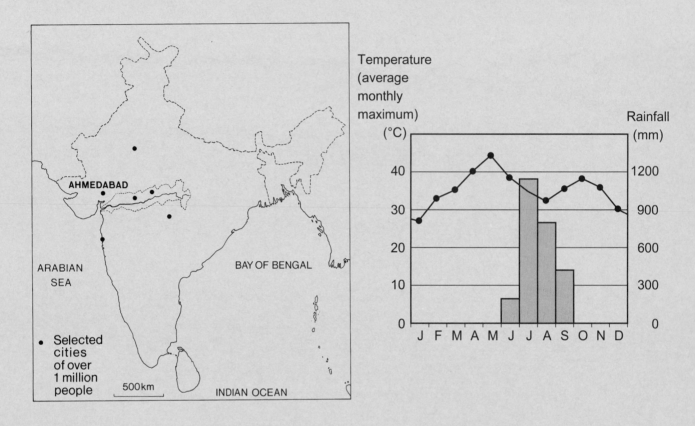

**Reference Map Q3E (Basin of Narmada River)**

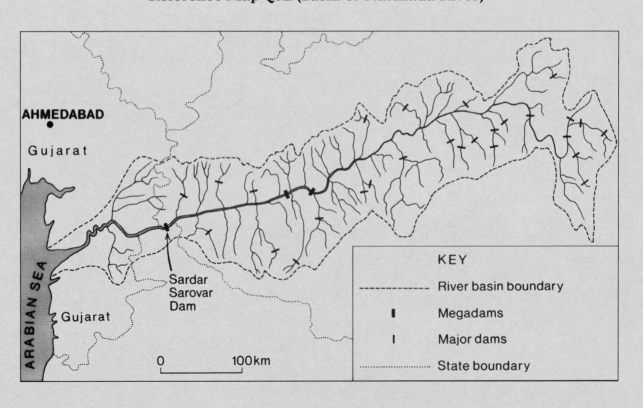

[Turn over for Question 4 on *Page ten*

*Marks*

## SECTION 2

**You must answer ONE question from this Section.**

**Question 4** (Urban Change and its Management)

Study Reference Maps Q4A and Q4B.

(a) **Describe** and **suggest reasons** for the changing distribution of the world's largest urban areas over the last 50 years.

**6**

(b) With reference to a named city which you have studied in an **ELDC** (Economically Less Developed Country):

   (i) **describe** the social, economic and environmental problems which have resulted from its rapid growth;

   (ii) **describe** some of the methods used to tackle these problems; and

   (iii) **comment** on the effectiveness of the methods used.

**10**

(c) The rural-urban fringe of many cities in **EMDCs** (Economically More Developed Countries) is frequently an area of land-use conflict.

Study Reference Map Q4C on *Page twelve*.

Referring to Edinburgh **or** any other named city which you have studied in an **EMDC**:

   (i) **suggest** why land-use conflicts may have arisen; and

   (ii) **comment** on the effectiveness of strategies such as the creation of Green Belts, in resolving these conflicts.

**9**

**(25)**

## Question 4 - continued

### Reference Map Q4A (Ten largest urban areas in the world in 1957)

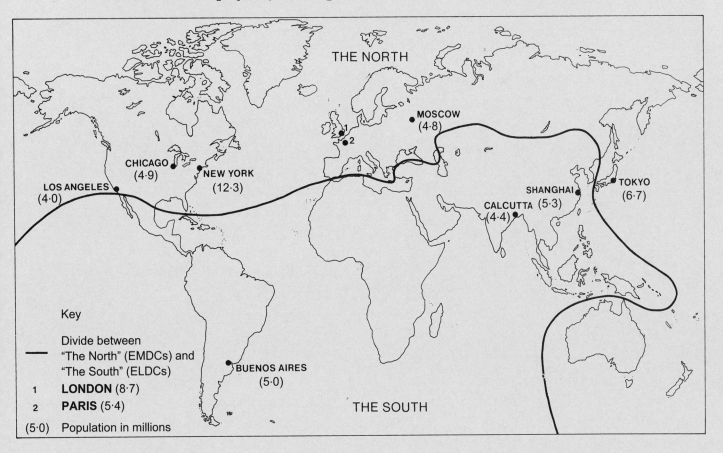

### Reference Map Q4B (Ten largest urban areas in the world in 2007)

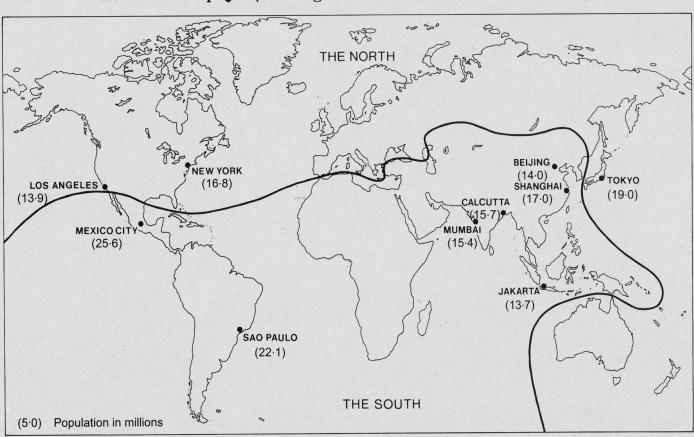

**Question 4 – continued**

Reference Map Q4C (Pressures on Edinburgh's Green Belt)

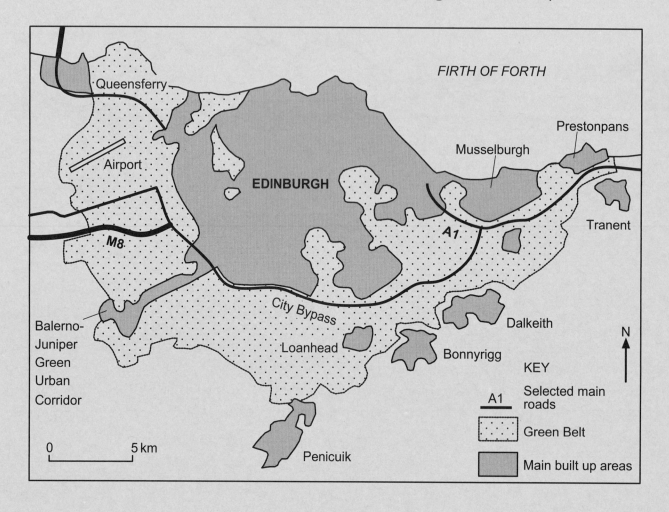

**[Turn over for Question 5 on *Page fourteen***

*Marks*

**Question 5** (European Regional Inequalities)

(a) Study Reference Table Q5A and Reference Map Q5A.

**Describe** and **suggest reasons** for the differences in levels of development between the countries shown.

5

(b) Study Reference Map Q5B on *Page sixteen* and Reference Table Q5B.

To what extent does the data provide evidence of regional inequalities within Poland?

6

(c) Many countries within the European Union (EU) have marked regional inequalities.

For a named country in the EU, select **one** less developed region and:

(i) **describe** and **account for** the social and economic problems faced by the region;

6

(ii) **outline** the efforts being made by both the national government and European Union agencies to tackle the problems, and comment on their effectiveness.

8

(25)

### Reference Table Q5A (Selected EU country statistics)

|  | France | Spain | Poland |
|---|---|---|---|
| Population (millions) | 60 | 41 | 39 |
| GDP ($US per capita) | 26 920 | 21 460 | 10 560 |
| Life expectancy | 79 | 79 | 74 |
| Health expenditure ($US per capita) | 2567 | 1607 | 629 |
| Urban % | 76 | 76 | 62 |
| Date of EU membership | 1957 | 1986 | 2004 |

Data: UN World Development Handbook

**Question 5 – continued**

### Reference Map Q5A (Selected European Union countries)

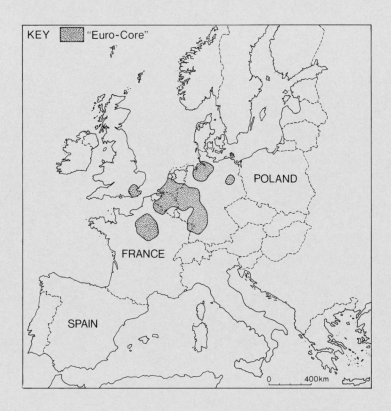

### Reference Table Q5B (Polish Provinces—selected statistics)

| Province | GDP per capita (in 1000 zloty) 2002 | Private vehicles % national total 2003 | Electricity production % total 2003 |
|---|---|---|---|
| Lower Silesia | 21·2 | 7 | 8 |
| Cuiavia & Pomerania | 18·6 | 5 | 2 |
| Lublin | 14·3 | 5 | 1 |
| Lubusz | 17·8 | 3 | <1 |
| Łódź | 18·5 | 7 | 20 |
| Lesser Poland | 17·7 | 8 | 6 |
| Mazovia | 31·1 | 15 | 13 |
| Opole | 16·7 | 3 | 6 |
| Sub-Carpathia | 14·6 | 5 | 2 |
| Podlassia | 15·7 | 3 | <1 |
| Pomerania | 20·3 | 6 | 2 |
| Silesia | 22·6 | 12 | 20 |
| Kielce | 16·0 | 3 | 5 |
| Varmia & Masuria | 15·2 | 3 | <1 |
| Greater Poland | 21·0 | 11 | 10 |
| Western Pomerania | 20·2 | 4 | 4 |

**[Turn over to see map of Provinces**

**Question 5 – continued**

### Reference Map Q5B (Poland: Provinces)

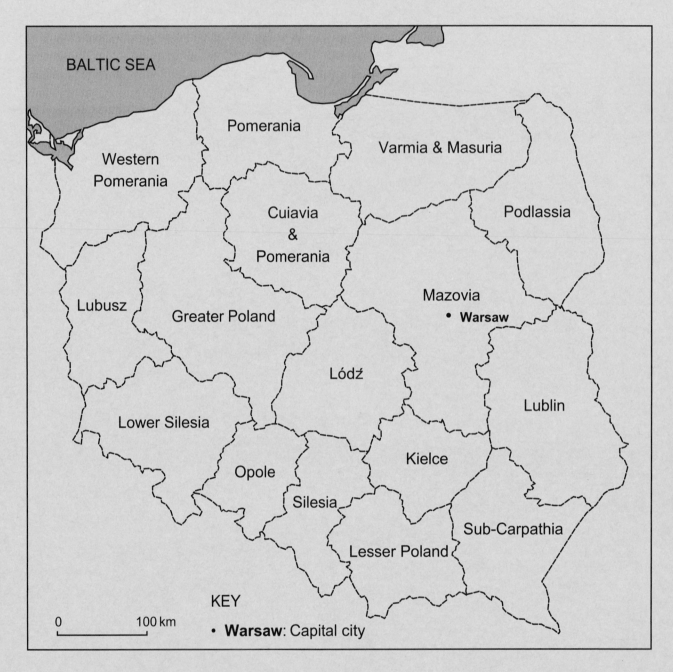

[Turn over for Question 6 on *Page eighteen*

*Marks*

**Question 6** (Development and Health)

(*a*) % Adult literacy is a social indicator of development. Identify one other social indicator of development **and** one economic indicator of development. For each indicator you have identified, **explain** how it might illustrate a country's level of development.

4

(*b*) Study Reference Table Q6.

**Reference Table Q6 (Adult literacy rates in selected Economically Less Developed Countries (ELDCs))**

| Country | % Adult Literacy |
| --- | --- |
| Afghanistan | 35 |
| Bolivia | 85 |
| Burkina Faso | 22 |
| Cuba | 96 |
| Kenya | 81 |
| Malaysia | 87 |
| Sri Lanka | 91 |

The table shows that there are considerable differences in levels of development between Economically Less Developed Countries (ELDCs). Referring to these countries and/or to other ELDCs you have studied, **suggest reasons** why such differences exist **between** countries.

5

(*c*) Many ELDCs have marked differences in levels of development **within** their borders. For a named ELDC, **explain** the differences found **within** the country.

4

(*d*) Study Reference Map Q6 which shows the main areas of the world at risk from cholera.

Referring to cholera **or** malaria **or** bilharzia/schistosomiasis:

(i) **describe** the physical and human factors which put people at risk of contracting the disease;

(ii) **describe** and **explain** the strategies used in controlling the spread of the disease; and

(iii) **explain** the benefits to ELDCs of controlling the disease.

12

**(25)**

**Question 6 – continued**

### Reference Map Q6 (Countries with a recent cholera outbreak)

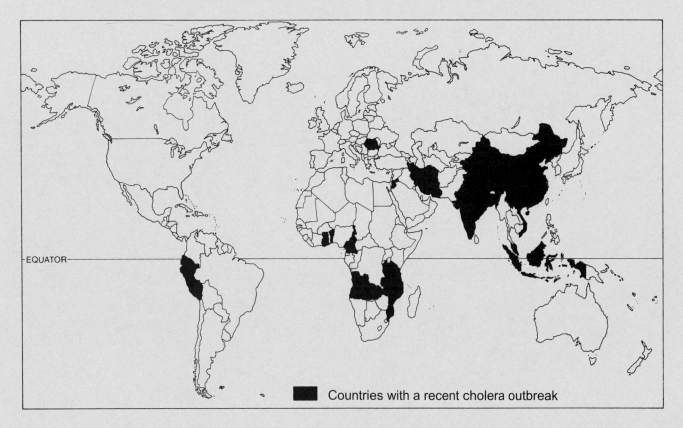

*[END OF QUESTION PAPER]*

[BLANK PAGE]

[BLANK PAGE]

# X208/301

| NATIONAL QUALIFICATIONS 2008 | THURSDAY, 22 MAY 9.00 AM – 10.30 AM | GEOGRAPHY HIGHER Paper 1 Physical and Human Environments |

**Six** questions should be attempted, namely:

**all four** questions in **Section A** (Questions 1, 2, 3 and 4);

**one** question from **Section B** (Question 5 **or** Question 6);

**one** question from **Section C** (Question 7 **or** Question 8).

Write the numbers of the **six** questions you have attempted in the marks grid on the back cover of your answer booklet.

The value attached to each question is shown in the margin.

Credit will be given for appropriate maps and diagrams, and for reference to named examples.

Questions should be answered in sentences.

**Note**    The reference maps and diagrams in this paper have been printed in black only: no other colours have been used.

Extract No 1659/EXP-OL15

1:25 000 Scale
Explorer Series

Four colours should appear above; if not then please return to the invigilator.
Four colours should appear above; if not then please return to the invigilator.

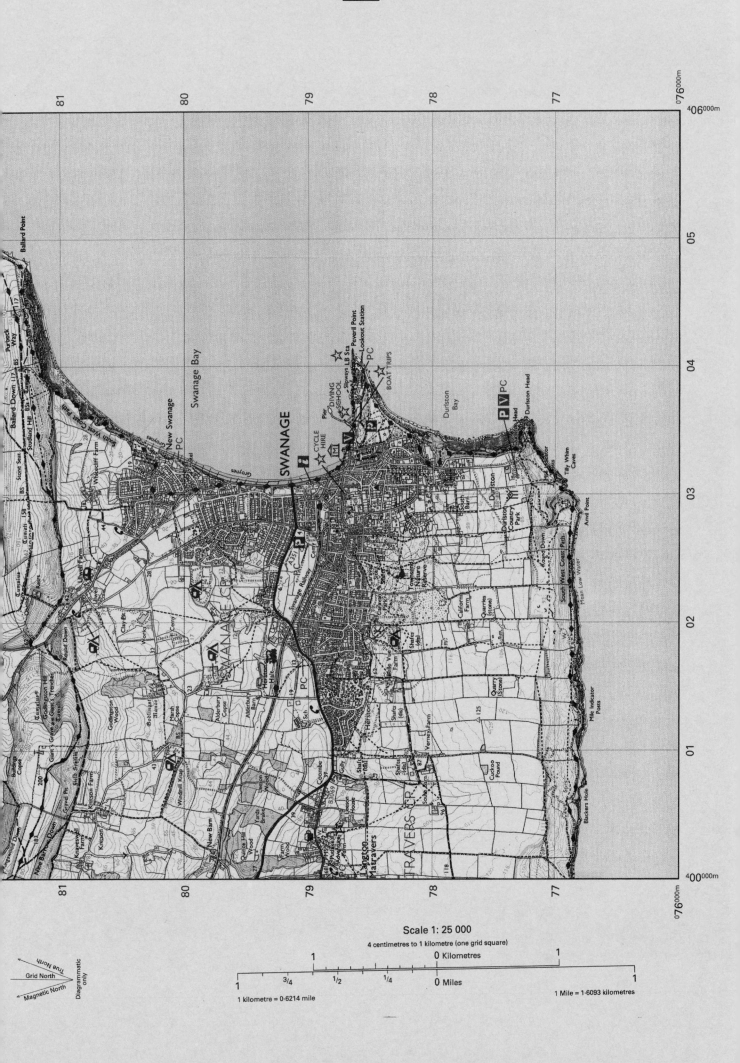

Scale 1: 25 000

4 centimetres to 1 kilometre (one grid square)

*Marks*

### SECTION A:  Answer ALL questions in this section

## Question 1:  Lithosphere

Study OS Map Extract number 1659/Exp–OL15: Swanage (*separate item*), **and** Reference Map Q1.

(*a*) **Describe** the map evidence that shows:

    (i)   Areas A and B are areas of coastal erosion, and

    (ii)  Area C is an area of coastal deposition.                    **12**

(*b*) With the aid of annotated diagrams, **explain** the various stages and processes involved in the formation of **either** a stack **or** a sand bar.                    **8**

### Reference Map Q1

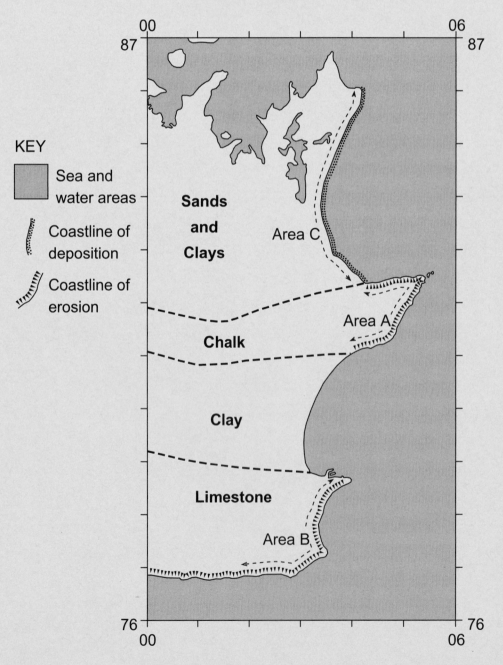

*Marks*

## Question 2:  Biosphere

Study Reference Diagram Q2.

**Describe** and **give reasons for** the changes in plant types likely to be observed across the transect as you move inland from the coast.

You should refer to named plant species likely to be found growing at different sites and to influencing factors **such as** shelter, pH and distance from the sea.

**16**

**Reference Diagram Q2 (Transect across sand dune coastline)**

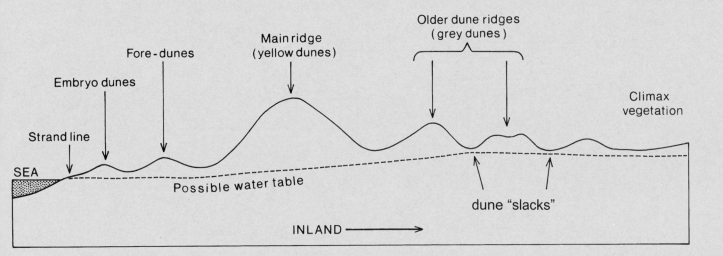

**[Turn over**

*Marks*

**Question 3: Population Geography**

Study Reference Diagram Q3 which shows the five stages of the Model of Demographic Transition.

(*a*) **Describe** and **explain** the changes in the **total population** in stages 1, 2 and 3 of the model.

**10**

(*b*) The total population levels off in stage 4 and starts to fall in stage 5.

**Describe** the problems which a government may face when a country is in stage 5.

**8**

**Reference Diagram Q3 (Model of demographic transition)**

| Stage 1 | Stage 2 | Stage 3 | Stage 4 | Stage 5 |
|---------|---------|---------|---------|---------|

*Marks*

## Question 4: Urban Geography

(*a*) For a named city which you have studied in an EMDC (Economically More Developed Country), **explain** the ways in which the site and situation have contributed to its growth.

8

(*b*) Study Reference Photograph Q4.

"*Traffic congestion is now a major problem facing many cities in EMDCs*".

**Describe** and **explain** schemes which have been introduced to reduce problems of traffic management in any named city you have studied in an EMDC.

10

**Reference Photograph Q4 (Traffic congestion)**

**[Turn over**

*Marks*

**SECTION B:  Answer ONE question from this section,
ie either Question 5 or Question 6.**

**Question 5:  Atmosphere**

Study Reference Diagram Q5.

**Explain** the physical **and** human factors that might have led to the changes in
global air temperatures shown in the diagram.

**14**

**Reference Diagram Q5 (Global air temperatures 1855–2005)**

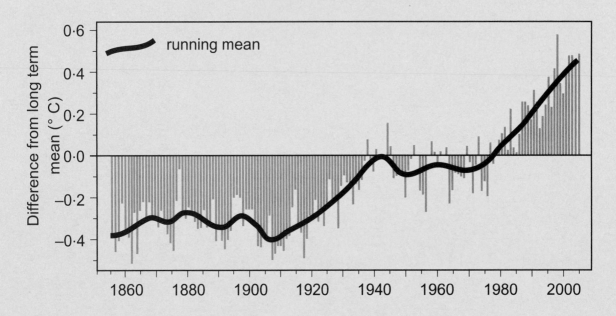

*Marks*

**DO NOT ANSWER THIS QUESTION IF YOU HAVE
ALREADY ANSWERED QUESTION 5**

## Question 6: Hydrosphere

(*a*)  With the aid of a diagram, **describe** the global hydrological cycle.     **6**

(*b*)  Study Reference Diagram Q6.

     **Explain** the **differences** in discharge between the urban and rural hydrographs shown in the diagram following a heavy rain storm.     **8**

**Reference Diagram Q6 (Flood hydrographs)**

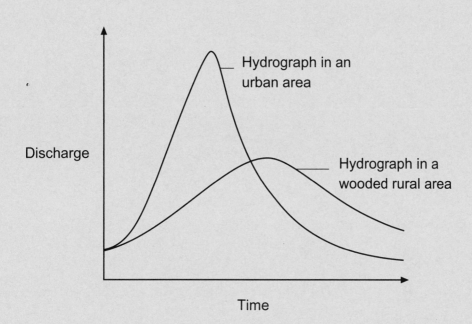

**[Turn over**

*Marks*

### SECTION C:  Answer ONE question from this section,
### ie either Question 7 or Question 8.

## Question 7:  Rural Geography

Study Reference Diagram Q7 which shows three different farming systems.

Choose **one** of these farming systems and:

(i)   **explain** the ways in which the diagram reflects the main features of your chosen system;                                                                                    **6**

(ii)  referring to a **named** area where **your chosen system** is carried out, **describe** the changes in farming practices that have taken place in recent years.              **8**

### Reference Diagram Q7 (Farming systems)

**Intensive peasant farming**        **Commercial arable farming**        **Shifting cultivation**

**DO NOT ANSWER THIS QUESTION IF YOU HAVE
ALREADY ANSWERED QUESTION 7**

*Marks*

## Question 8: Industrial Geography

(*a*) Study Reference Diagram Q8A.

**Describe** and **explain** the impact of industry on the environment of an old industrial area such as that shown in Reference Diagram Q8A.

**6**

**Reference Diagram Q8A (Old industrial landscape—South Wales)**

(*b*) Study Reference Diagram Q8B.

For South Wales, or any other industrial concentration in the EU, **describe** and **explain** the main location factors that influence the location of new industrial developments.

**8**

**Reference Diagram Q8B (New industrial landscape—South Wales)**

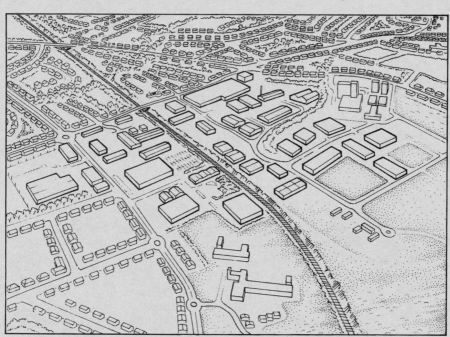

*[END OF QUESTION PAPER]*

[BLANK PAGE]

# X208/303

NATIONAL
QUALIFICATIONS
2008

THURSDAY, 22 MAY
10.50 AM – 12.05 PM

GEOGRAPHY
HIGHER
Paper 2
Environmental
Interactions

**Two** questions should be attempted, namely:

**one** question from **Section 1** (Questions 1, 2, 3) and
**one** question from **Section 2** (Questions 4, 5, 6).

Write the numbers of the **two** questions you have attempted in the marks grid on the back cover of your answer booklet.

The value attached to each question is shown in the margin.

Credit will be given for appropriate maps and diagrams, and for reference to named examples.

Questions should be answered in sentences.

**Note**   The reference maps and diagrams in this paper have been printed in black only: no other colours have been used.

*Marks*

## SECTION 1

**You must answer ONE question from this Section.**

**Question 1** (Rural Land Resources)

(*a*) The Peak District and Yorkshire Dales National Parks are two areas of Upland Limestone.

Study Reference Diagram Q1A.

**Describe** and **explain** the physical features associated with upland limestone landscapes. Both surface and underground features should be included in your answer.                                                                 **20**

(*b*) For the Peak District National Park, **or** a named upland area you have studied:

  (i) **describe** the opportunities which this landscape provides for a variety of land uses; and                                                    **8**

  (ii) **explain** the environmental problems and conflicts which may arise from the competing demands of these different land uses.            **14**

(*c*) Study Reference Diagram Q1B.

Select **one** of the conservation strategies and **explain** the ways in which it helps to protect the landscape.                                        **8**

**(50)**

**Question 1 – continued**

**Reference Diagram Q1A (Carboniferous Limestone Landscape)**

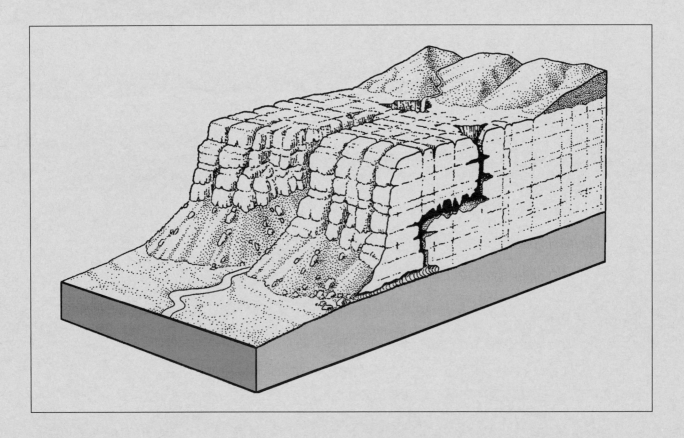

**Reference Diagram Q1B (Conservation Strategies)**

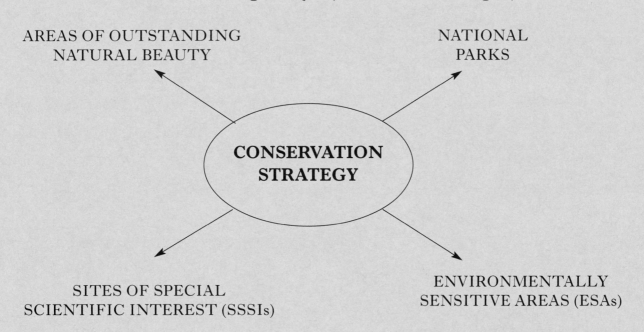

AREAS OF OUTSTANDING
NATURAL BEAUTY

NATIONAL
PARKS

**CONSERVATION
STRATEGY**

SITES OF SPECIAL
SCIENTIFIC INTEREST (SSSIs)

ENVIRONMENTALLY
SENSITIVE AREAS (ESAs)

**[Turn over**

*Marks*

**Question 2** (Rural Land Degradation)

(a) Study Reference Map Q2A and Reference Maps Q2B.

   **Describe** the climatic conditions found in Burkina Faso, and **explain** why such conditions may lead to the degradation of rural land.    **16**

(b) **Explain** how inappropriate farming activities such as overcultivation, monoculture, overgrazing, poor irrigation techniques and inappropriate cultivation of marginal land have led to land degradation in some named areas of North America.    **8**

(c) Study Reference Statements Q2A and Q2B.

   Select **one** of the statements and **explain** how degradation has impacted on the social and economic ways of life in that area.    **10**

(d) Referring to named areas in North America that you have studied, **describe** and **explain** ways in which changes in farming methods have reduced land degradation.    **16**

   **(50)**

### Reference Statement Q2A (The Sahel)

> **28·5 MILLION PEOPLE ARE AFFECTED BY DESERTIFICATION IN THE SAHEL REGION OF AFRICA.**

### Reference Statement Q2B (The Amazon Basin)

> **179 000 SQUARE KILOMETRES OF RAINFOREST HAVE DISAPPEARED IN THE AMAZON BASIN SINCE 1997, EQUIVALENT TO 78% OF THE TOTAL AREA OF BRITAIN.**

**Question 2 – continued**

## Reference Map Q2A (Climatic Regions of Burkina Faso)

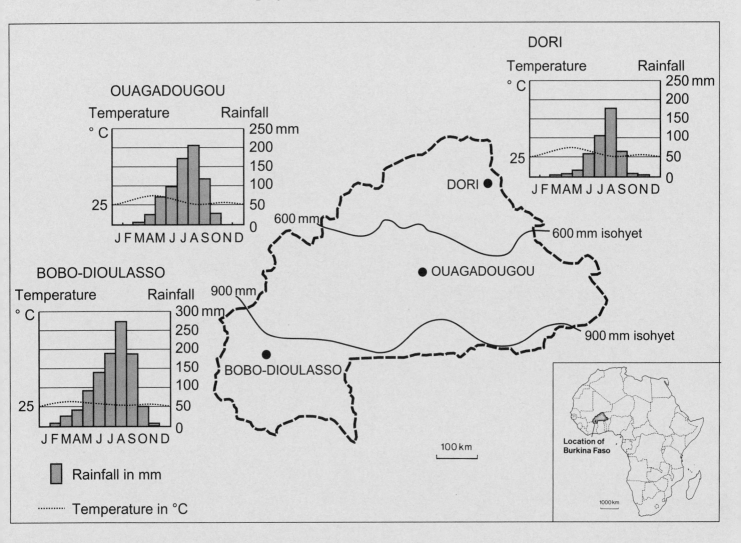

## Reference Maps Q2B (Burkina Faso: mean annual rainfall patterns)

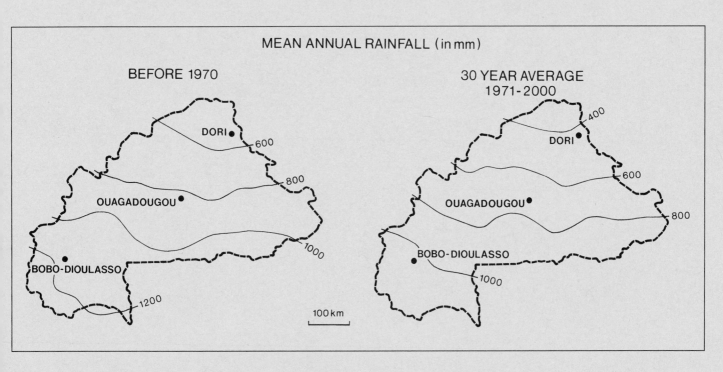

**Question 3** (River Basin Management)

(a)  Study Reference Map Q3 and Reference Diagram Q3.

   **Explain** why there is a need for water management in Egypt.    **10**

(b)  For the Aswan High Dam **or** any dam you have studied in Africa **or** North America **or** Asia, **explain** the **physical** factors which should be considered when selecting the site for the dam and associated reservoir.    **10**

(c)  **Describe** and **explain** the social, economic and environmental benefits **and** adverse consequences of a named major water control project in Africa **or** North America **or** Asia.    **24**

(d)  "*Potential 'water wars' are likely in areas where rivers and lakes are shared by more than one country or state, according to a UN Development Programme (UNDP) report.*"

   **Explain** why political problems can occur in the development of water control projects.    **6**

   **(50)**

## Reference Map Q3 (The Nile Basin)

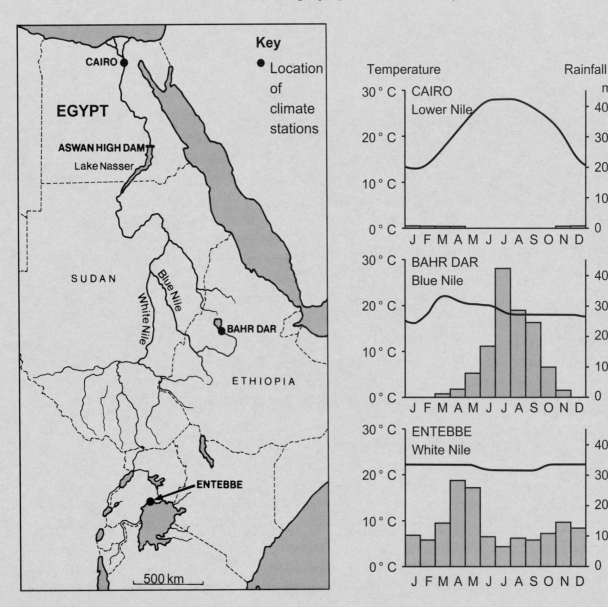

**Question 3 – continued**

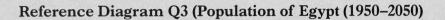

### Reference Diagram Q3 (Population of Egypt (1950–2050)

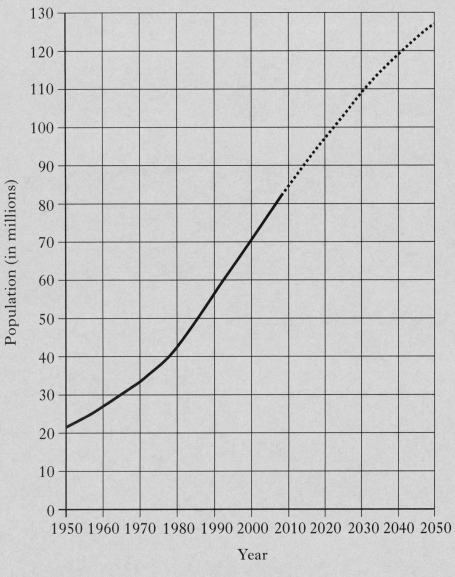

actual figures

estimate

[**Turn over**

*Marks*

## SECTION 2

### You must answer ONE question from this Section.

**Question 4** (Urban Change and its Management)

(*a*) Study Reference Table Q4A.

(i) **Describe** the changes shown in the table.

(ii) **Suggest** reasons for the differences between more developed and less developed regions.	**14**

(*b*) With the aid of Reference Photograph Q4 and referring to a named city which you have studied in an **ELDC (Economically Less Developed Country)**:

(i) **describe** the social, economic and environmental problems created by shanty towns; and

(ii) **describe ways** in which such problems are being tackled.	**18**

(*c*) Study Reference Table Q4B which highlights problems which have occurred in cities in **EMDCs (Economically More Developed Countries)** over the last fifty years. Choose **one** of these problems and, with reference to a named city in an EMDC:

(i) **suggest** reasons for the problem;

(ii) **describe** strategies used to solve the problem; and

(iii) **comment** on the success of these strategies.	**18**
	**(50)**

**Question 4 – continued**

Reference Table Q4A (Percentage of total population living in urban areas)

| Region | Urban population (%) | | |
|---|---|---|---|
| | 1970 | 1994 | 2025 |
| **More developed regions** | **67** | **75** | **84** |
| Europe | 64 | 73 | 83 |
| North America | 74 | 76 | 85 |
| **Less developed regions** | **25** | **37** | **57** |
| Africa | 23 | 33 | 54 |
| South and Central America | 57 | 74 | 85 |

Reference Photograph Q4 (A Shanty Town in Cape Town, South Africa)

Reference Table Q4B (Selected problems facing cities in EMDCs
(Economically More Developed Countries))

- **Housing change in the Inner City**
- **The decline of traditional industries**
- **The rise of out of town shopping**

*Marks*

**Question 5** (European Regional Inequalities)

(*a*)  Study Reference Map Q5.

The European Union (EU) is often said to fit the "Core and Periphery" model.  Ten of the twelve countries which joined the EU since 2004 have formed a new "Eastern Periphery".

**Suggest** both physical **and** human reasons for the lack of prosperity in the new "Eastern Periphery".                                                                       **10**

(*b*)  Study Reference Table Q5 which shows a range of indicators for six European Union countries.

**Describe** and **explain** the ways in which the information shows the **differences** between the three **groups** of countries shown in the table.    **12**

(*c*)  "There are marked differences in economic development within the United Kingdom (UK)."

**Describe** and **explain** both the physical **and** human factors that have led to regional inequalities within the UK.                                          **14**

(*d*)  For **either** the UK **or** another named country in the EU which has marked regional differences in economic development, **discuss** ways in which the National Government **and** the EU have tried to tackle problems in less prosperous regions.                                                                                      **14**

**(50)**

**Question 5 – continued**

**Reference Map Q5 (The Core and Eastern Periphery of the European Union)**

New Eastern Periphery countries

Euro Core

**Reference Table Q5 (Selected indicators of development)**

| Economic group | Country | Infant mortality rate (per 1000 live births) | GDP per capita ($) | Employment (%) | | |
|---|---|---|---|---|---|---|
| | | | | Primary | Secondary | Tertiary |
| Euro-Core Countries | Germany | 4 | 30 400 | 2 | 26 | 72 |
| | Netherlands | 5 | 30 500 | 2 | 19 | 79 |
| Pre-2004 Periphery Countries | Greece | 5 | 22 200 | 12 | 20 | 68 |
| | Portugal | 5 | 19 300 | 10 | 30 | 60 |
| Eastern Periphery Countries | Estonia | 8 | 16 700 | 11 | 20 | 69 |
| | Slovakia | 7 | 16 000 | 6 | 29 | 65 |

*Marks*

**Question 6** (Development and Health)

(a)  Study Reference Map Q6.

(i)  **Describe** clearly **two** economic and **two** social indicators of development which could be used to produce a map such as this.    **8**

(ii)  **Suggest reasons** for the wide variations in development which exist **between** Economically Less Developed Countries (ELDCs).

You should refer to named ELDCs you have studied.    **12**

(iii)  There are often considerable differences in levels of development and living standards **within** a single country.

Referring to a named ELDC which you have studied, **suggest reasons** why such regional variations exist.    **10**

(b)  For **either** malaria **or** bilharzia **or** cholera:

(i)  **describe** the environmental **and** human factors which put people at risk of contracting the disease; and    **8**

(ii)  **describe** and **evaluate** the methods used to control the spread of the disease.    **12**

**(50)**

**Question 6 – continued**

**Reference Map Q6 (The World:  Human Development Index (HDI))**

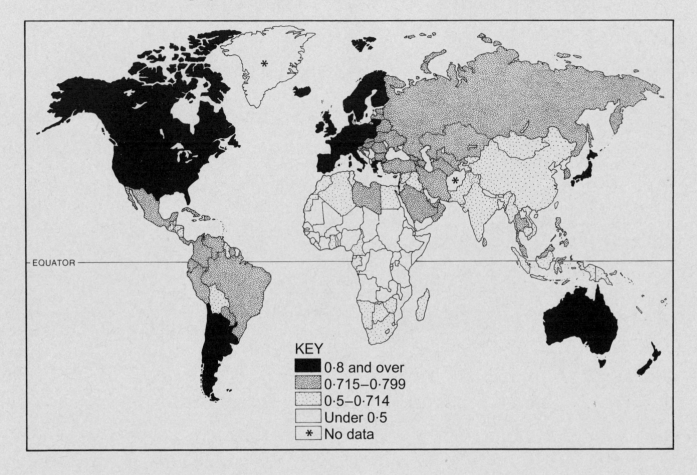

KEY

- ■ 0·8 and over
- ▨ 0·715–0·799
- ▨ 0·5–0·714
- ☐ Under 0·5
- * No data

[END OF QUESTION PAPER]

[BLANK PAGE]

[BLANK PAGE]

# X208/301

| | | |
|---|---|---|
| NATIONAL QUALIFICATIONS 2009 | WEDNESDAY, 27 MAY 9.00 AM – 10.30 AM | GEOGRAPHY HIGHER Paper 1 Physical and Human Environments |

**Six** questions should be attempted, namely:

**all four** questions in **Section A** (Questions 1, 2, 3 and 4);

**one** question from **Section B** (Question 5 **or** Question 6);

**one** question from **Section C** (Question 7 **or** Question 8).

Write the numbers of the **six** questions you have attempted in the marks grid on the back cover of your answer booklet.

The value attached to each question is shown in the margin.

Credit will be given for appropriate maps and diagrams, and for reference to named examples.

Questions should be answered in sentences.

**Note**    The reference maps and diagrams in this paper have been printed in black only: no other colours have been used.

1:50 000 Scale
Landranger Series

*Marks*

### SECTION A: Answer ALL questions in this section

**Question 1: Hydrosphere**

Study OS Map Extract number 1745/98: Upper Wharfedale (*separate item*).

(*a*)  Using appropriate grid references, **describe** the physical characteristics of the River Wharfe and its valley from 978690 to 040603.

**10**

(*b*)  **Explain**, with the aid of a diagram or diagrams, how a waterfall is formed in the upper course of a river valley.

**8**

*Marks*

## Question 2:  Biosphere

(*a*) **Draw** and **fully annotate** a soil profile of a **podzol** to show its main characteristics (including horizons, colour, texture and drainage) and associated vegetation.     **9**

Study Reference Diagram Q2 which shows a soil profile of a brown earth soil.

(*b*) **Describe** and **explain** the formation and characteristics of a **brown earth soil**.     **9**

### Reference Diagram Q2

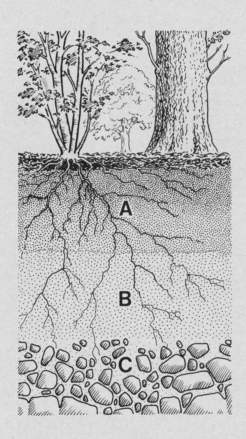

**[Turn over**

*Marks*

**Question 3: Rural Geography**

(*a*)  Study Reference Table Q3.

**Describe** and **explain**, with the aid of the data in the table, the differences between intensive peasant farming and commercial arable farming.

**10**

**Reference Table Q3 (Types of farming and selected data)**

|  | **Bangladesh** | **Canada** |
|---|---|---|
| **Farm type** | Intensive peasant farming | Commercial arable farming |
| **GDP per capita ($US)** | 158 | 13 034 |
| **% GDP from farming** | 48 | 3 |
| **% population engaged in farming** | 82 | 4 |
| **Kg of fertiliser used per hectare** | 26 | 32 |
| **People per tractor** | 20 581 | 38 |

(*b*)  Areas of intensive peasant farming such as those in Bangladesh have undergone changes in recent years.

Referring to an area you have studied:

 (i) **describe** these changes, and

 (ii) **outline** the impact of these changes on the people **and** the farming landscape.

**10**

*Marks*

## Question 4:  Industrial Geography

"*Many industrial concentrations within the European Union have undergone a great transformation in the last 50 years.  These changes are most marked in the types of industries, the industrial landscape and in employment patterns.*"

Referring to a **named** industrial concentration in the European Union that you have studied:

(i)  **describe** and **account for** the main characteristics of a typical "new" industrial landscape;                                                                 9

(ii)  **suggest** ways in which the national government **and** the European Union have helped to attract new industries to your chosen area.                          7

**[Turn over**

*Marks*

**SECTION B:  Answer ONE question from this section,
ie either Question 5 or Question 6.**

**Question 5:  Atmosphere**

*"Energy is transferred from areas of surplus, between 35 °N and 35 °S, to areas of deficit, polewards from 35 °N and 35 °S, by both oceanic and atmospheric circulation."*

Study Reference Map Q5 which shows selected ocean currents in the North Atlantic Ocean.

(*a*)  (i)  **Describe** the pattern of ocean currents in the North Atlantic Ocean, and

(ii)  **explain** how they help to maintain the global energy balance.

6

**Reference Map Q5 (Selected ocean currents in the North Atlantic Ocean)**

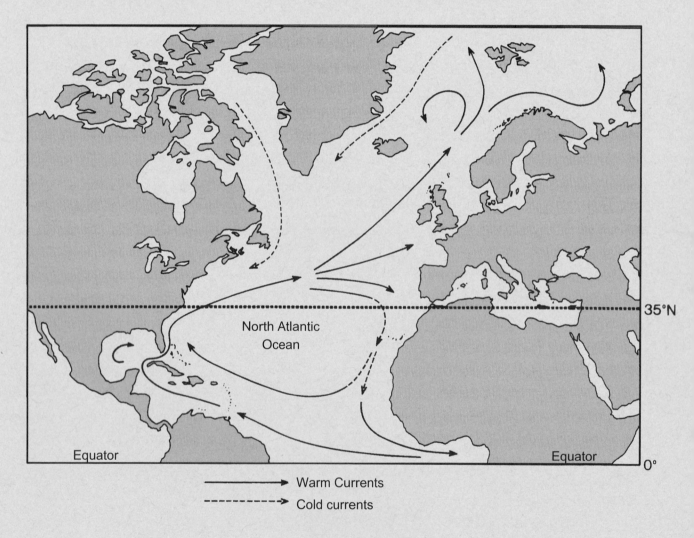

⎯⎯⎯⎯→ Warm Currents

---------→ Cold currents

*Marks*

## Question 5:  Atmosphere (continued)

Study Reference Diagram Q5 which shows surface winds and pressure zones.

(*b*) **Explain** how circulation cells in the atmosphere and the associated surface winds assist in the transfer of energy between areas of surplus and deficit.    **8**

### Reference Diagram Q5 (Surface winds and pressure zones)

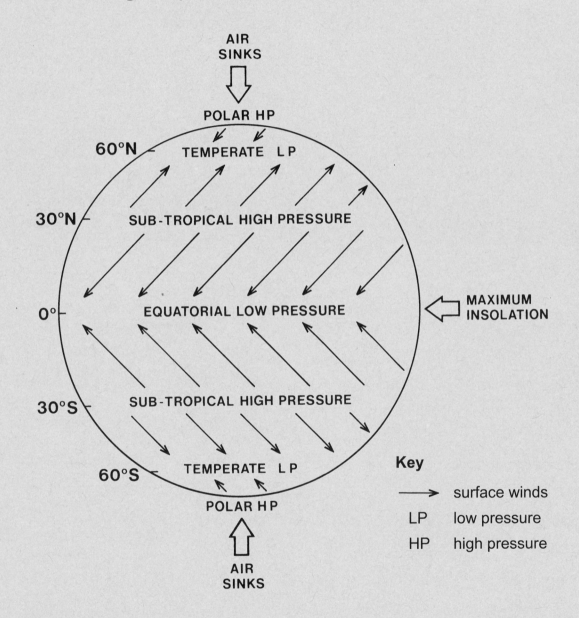

[**Turn over**

*Marks*

### DO NOT ANSWER THIS QUESTION IF YOU HAVE
### ALREADY ANSWERED QUESTION 5

**Question 6: Lithosphere**

Study OS Map Extract number 1745/98: Upper Wharfedale (*separate item*), and Reference Map Q6.

*The map extract covers part of the Yorkshire Dales National Park, an area famed for its Carboniferous Limestone scenery, characterised by distinctive surface features, drainage patterns and underground landforms.*

(*a*) **Describe** the evidence which suggests that Area A, shown on Reference Map Q6, is a Carboniferous Limestone landscape.
(You should refer to named features and make use of grid references.)    8

(*b*) Choose any **one** Carboniferous Limestone feature described in your answer to part (*a*) and, with the aid of annotated diagrams, **explain** how it was formed.    6

### Reference Map Q6

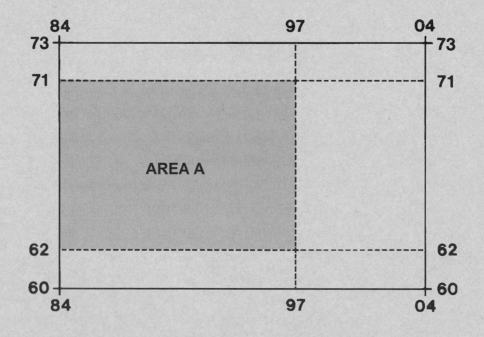

*Marks*

**SECTION C: Answer ONE question from this section,**
**ie either Question 7 or Question 8.**

### Question 7: Population

Italy has a population structure that is typical of many EMDCs (Economically More Developed Countries).

Study Reference Diagrams Q7A and Q7B.

(a) **Describe** and **account** for the changes between the population structure in 2000 and that projected for 2050.

8

(b) **Discuss** the consequences of the 2050 population structure for the future economy of the country and the welfare of its citizens.

6

**Reference Diagram Q7A (Italy: Population pyramid for 2000)**

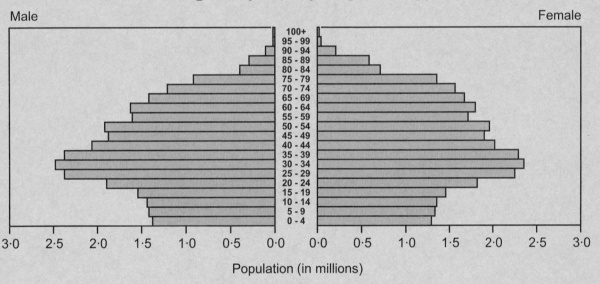

**Reference Diagram Q7B (Italy: Population pyramid for 2050)**

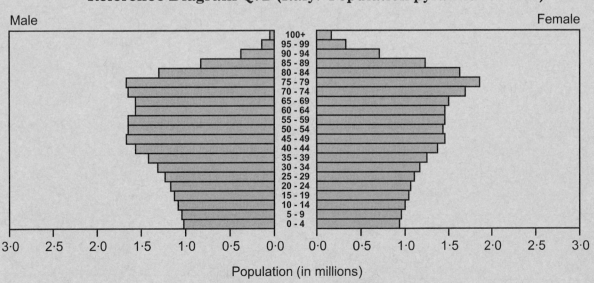

*Marks*

**DO NOT ANSWER THIS QUESTION IF YOU HAVE
ALREADY ANSWERED QUESTION 7**

**Question 8:  Urban Geography**

Study Reference Photograph Q8A which shows Buchanan Galleries shopping centre in Glasgow's CBD and Reference Photograph Q8B which shows Braehead, an out-of-town shopping centre situated at the south-west edge of Glasgow.

Referring to Glasgow, **or** any other named city you have studied in an Economically More Developed Country (EMDC):

(i)  **suggest** the impact that an out-of-town shopping centre may have had on shopping in the traditional CBD;                                                                      **6**

(ii)  **describe** and **explain** the changes, other than shopping, which have taken place in the CBD over the past few decades.                                                   **8**

**Reference Photograph Q8A**          **Reference Photograph Q8B**

*[END OF QUESTION PAPER]*

# X208/303

NATIONAL
QUALIFICATIONS
2009

WEDNESDAY, 27 MAY
10.50 AM – 12.05 PM

GEOGRAPHY
HIGHER
Paper 2
Environmental
Interactions

Answer any **two** questions.

Write the numbers of the **two** questions you have attempted in the marks grid on the back cover of your answer booklet.

The value attached to each question is shown in the margin.

Credit will be given for appropriate maps and diagrams, and for reference to named examples.

Questions should be answered in sentences.

**Note**    The reference maps and diagrams in this paper have been printed in black only: no other colours have been used.

*Marks*

**Question 1** (Rural Land Resources)

*Loch Lomond and the Trossachs became Scotland's first National Park in 2002. It covers 1865 square kilometres of lowland, river, loch, forest and mountain landscapes.*

(*a*) **Describe** and **explain**, with the aid of annotated diagrams, the formation of the main glacial features of the Loch Lomond and the Trossachs National Park **or** any other glaciated upland area in the UK that you have studied.

**20**

(*b*) With reference to Loch Lomond and the Trossachs **or** any other named upland area that you have studied, **explain** the social and economic opportunities created by the landscape.

**10**

(*c*) Study Reference Diagram Q1.

Reference Diagram Q1 shows the Loch Lomond and the Trossachs National Park to be under intense environmental pressure in certain key areas. With reference to this area or any **named** upland area you have studied:

(i) **describe** and **explain** the environmental conflicts that may occur (you should refer to named locations within your chosen upland landscape);

**10**

(ii) **describe** specific solutions to these environmental conflicts commenting on their effectiveness.

**10**

**(50)**

**Question 1 – continued**

**Reference Diagram Q1 (Loch Lomond and the Trossachs: Environmental Activity and Pressure)**

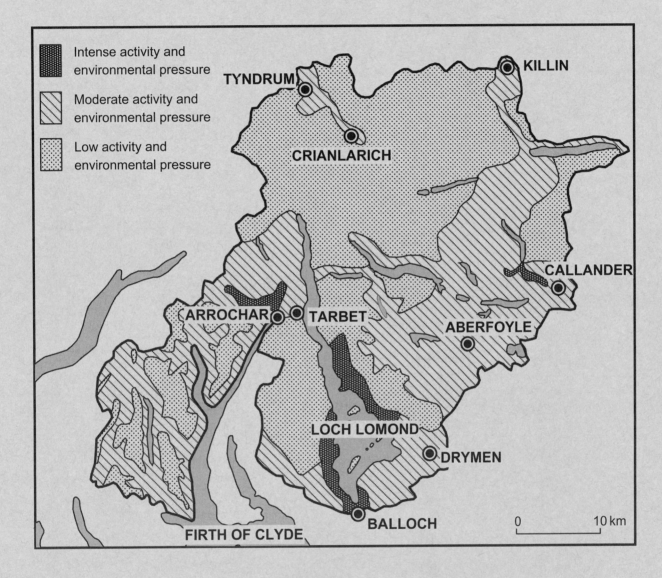

[Turn over

*Marks*

**Question 2** (Rural Land Degradation)

*The Sahel is a 500 kilometre wide zone which runs across Africa along the southern edge of the Sahara Desert. The Sahel is under intense pressure from human activity which, combined with climate change, has created a "spiral of desertification".*

(a)  Study Reference Diagram Q2.

**Describe** the changes in rainfall patterns shown on Reference Diagram Q2.    **6**

(b)  For **either** Africa north of the Equator **or** the Amazon Basin:

(i)  **explain** how human activities, including inappropriate farming techniques, have contributed to land degradation; and    **18**

(ii)  **describe** some of the consequences of land degradation on the people and their environment.    **10**

(c)  Referring to **named** areas of **North America** which you have studied:

(i)  **describe** some of the measures which have been taken to conserve soil and limit land degradation; and

(ii)  **comment on** the effectiveness of these measures.    **16**

**(50)**

**Question 2 – continued**

**Reference Diagram Q2 (Rainfall Variability in the Sahel)**

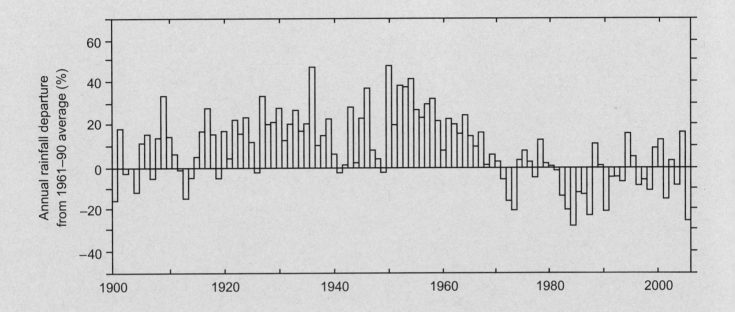

**[Turn over**

*Marks*

**Question 3** (River Basin Management)

(a)  Study Reference Table Q3 and Reference Map Q3.

Explain the need for water management in the Colorado Basin.                    **10**

(b)  Explain the physical **and** human factors that have to be considered when selecting sites for dams and their associated reservoirs.                    **14**

(c)  Study Reference Diagram Q3 and Reference Map Q3.

For the Colorado River Basin, **or** another river basin in North America, **or** in Africa, **or** in Asia, that you have studied:

(i)  describe the problems caused by the river flowing through more than one state or country;

(ii)  suggest ways in which these problems may be overcome.                    **10**

(d)  Describe and explain the social, economic and environmental benefits of a named water control project in North America or Africa or Asia.                    **16**

                    **(50)**

### Reference Table Q3 (Population Growth in Las Vegas and Phoenix)

| Selected city | 1990 Population | 2000 Population | Population change (1990–2000) |
|---|---|---|---|
| Phoenix | 2 238 480 | 3 251 876 | +45% |
| Las Vegas | 741 459 | 1 375 765 | +85% |

### Reference Diagram Q3 (The Colorado River Water Allocation)

**Upper Basin—Water allocation**          **Lower Basin—Water allocation**

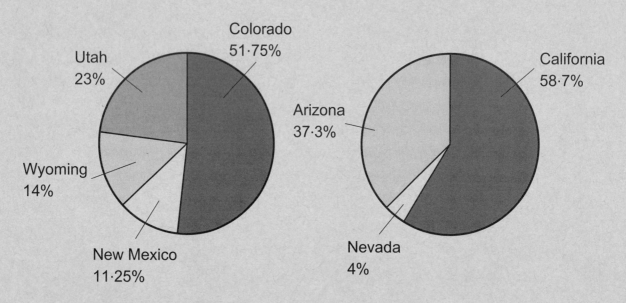

**Question 3 – continued**

## Reference Map Q3 (The Colorado River Basin)

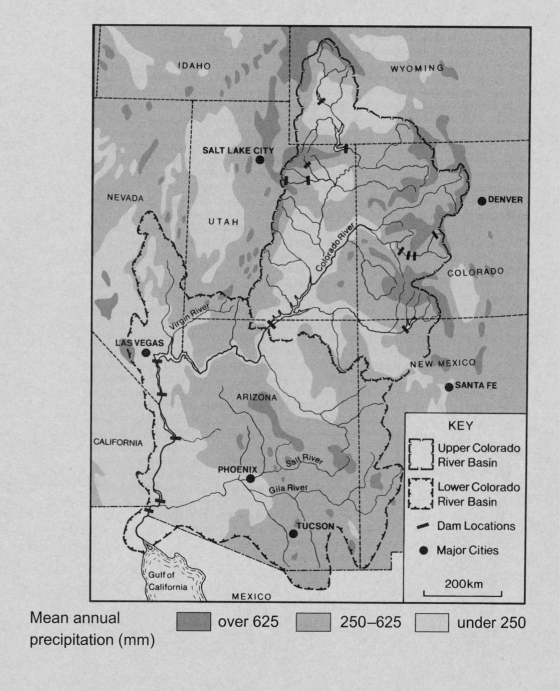

Mean annual precipitation (mm)    over 625    250–625    under 250

**[Turn over**

*Marks*

**Question 4** (Urban Change and its Management)

(*a*)  Study Reference Map Q4A.

**Describe** and **account for** the distribution of major cities in **either** Spain **or** any other EMDC (Economically More Developed Country) that you have studied.

**10**

(*b*)  "*Kibera is one of almost 100 shanty towns in Nairobi, the capital city of Kenya. More than half of Nairobi's 3 million people live in these shanties, which in total occupy less than 2% of the city's land area.*"

With reference to a named city that you have studied in an ELDC (Economically Less Developed Country):

(i)  **describe** the social, economic and environmental problems often found in these shanty town areas;

**12**

(ii)  **describe** the methods the shanty dwellers and the city authorities might use to tackle these problems, and comment on the effectiveness of these methods.

**8**

(*c*)  Study Reference Map Q4B.

The map shows the Aberdeen Western Peripheral Route (AWPR), a proposed new road to improve traffic management in and around Aberdeen and the North-east of Scotland.

For Aberdeen, or a **named** city that you have studied in an EMDC:

(i)  **describe** and **explain** why it suffers from traffic congestion;

**12**

(ii)  **suggest** why the building of major new roads such as the AWPR may lead to protests and land-use conflicts.

**8**

**(50)**

**Question 4 – continued**

### Reference Map Q4A (Largest Cities in Spain)

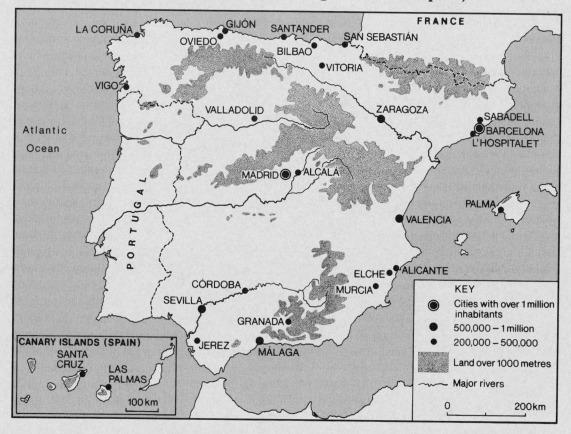

### Reference Map Q4B (Aberdeen Western Peripheral Route (AWPR))

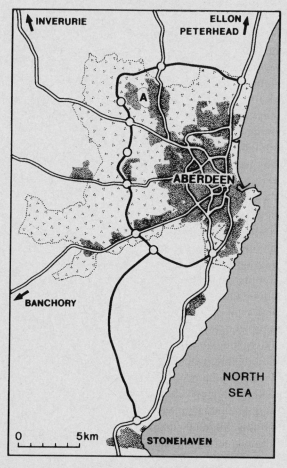

**Question 5** (European Regional Inequalities)

(*a*)   Study Reference Table Q5.

**Describe** and **suggest reasons** for the differences in levels of development between the pre-2000 EU member states and the post-2000 EU member states.     **10**

(*b*)   Study Reference Map Q5.

    (i) **Describe** the distribution of the regions which were eligible for European grants under Objective 1 support (2000–2006).     **8**

    (ii) **Explain** how EU initiatives such as Objective 1 support might improve the less prosperous regions of the European Union.     **8**

(*c*)   "The European Cohesion policy (2007–2013) aims to contribute towards economic and social cohesion within the EU by reducing regional differences and human inequality within member states."

For any named country you have studied in the European Union:

    (i) **describe** the physical and human factors which have led to regional inequalities;     **18**

    (ii) **outline** the steps taken by the national government to tackle these regional inequalities.     **6**

                 **(50)**

**Reference Map Q5 (European Union Objective 1 Funding)**

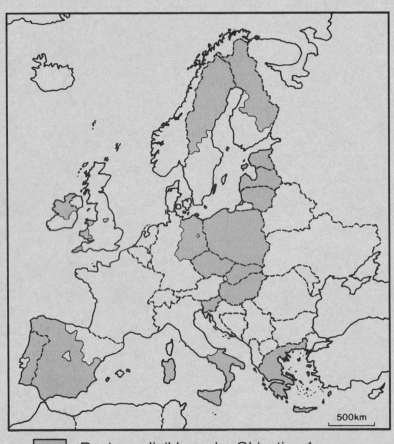

☐ Regions eligible under Objective 1

(Objective 1: Supporting development in less prosperous regions)

## Question 5 – continued

### Reference Table Q5 (European Union Statistics Ranked in Order)

| Pre–2000 Member States | | | | Post–2000 Member States | | | |
|---|---|---|---|---|---|---|---|
| Country | Year of EU membership | GDP (ranked)* | HDI (ranked)* | Country | Year of EU membership | GDP (ranked)* | HDI (ranked)* |
| Belgium | 1957 | 6 | 6 | Cyprus | 2004 | 14 | 17 |
| France | 1957 | 11 | 9 | Czech Rep | 2004 | 17 | 18 |
| Germany | 1957 | 10 | 13 | Estonia | 2004 | 20 | 22 |
| Italy | 1957 | 12 | 10 | Hungary | 2004 | 21 | 20 |
| Luxembourg | 1957 | 1 | 5 | Latvia | 2004 | 24 | 25 |
| Netherlands | 1957 | 3 | 3 | Lithuania | 2004 | 23 | 23 |
| Denmark | 1973 | 5 | 8 | Malta | 2004 | 18 | 19 |
| Ireland | 1973 | 2 | 1 | Poland | 2004 | 25 | 21 |
| UK | 1973 | 8 | 11 | Slovakia | 2004 | 22 | 24 |
| Greece | 1981 | 15 | 14 | Slovenia | 2004 | 16 | 15 |
| Portugal | 1986 | 19 | 16 | Bulgaria | 2007 | 26 | 26 |
| Spain | 1986 | 13 | 12 | Romania | 2007 | 27 | 27 |
| Finland | 1995 | 9 | 4 | | | | |
| Sweden | 1995 | 7 | 2 | | | | |
| Austria | 1995 | 4 | 7 | | | | |

GDP      Gross Domestic Product per capita reflects total of all goods and services per head of population

HDI      Human Development Index (covering poverty, education, health)

*Ranking    1–27 with 1 best and 27 worst

**[Turn over**

*Marks*

**Question 6** (Development and Health)

(a)  Study Reference Map Q6 which shows the Human Development Index (HDI) for countries of the world.

**Explain** the advantages of using a composite indicator of development such as the HDI rather than a single indicator.  **4**

(b)  Referring to named examples, **suggest reasons** why there is such a wide range in levels of development **between** different ELDCs (Economically Less Developed Countries).  **12**

(c)  For malaria, **or** bilharzia, **or** cholera:

  (i)  **describe** the human and environmental factors that can contribute to the spread of the disease;  **6**

  (ii)  **describe** the measures that have been taken to combat the disease;  **12**

  (iii)  **explain** how the eradication or control of the disease would benefit ELDCs.  **6**

(d)  *"Resources need to be targeted at improving Primary Health Care if we are ever going to improve the health of people in ELDCs."*  Aid worker

**Describe** some of the strategies involved in Primary Health Care and **explain** why these strategies for improving health standards are suited to people living in ELDCs.  **10**

**(50)**

**Question 6 – continued**

**Reference Map Q6 (The World:  Human Development Index)**

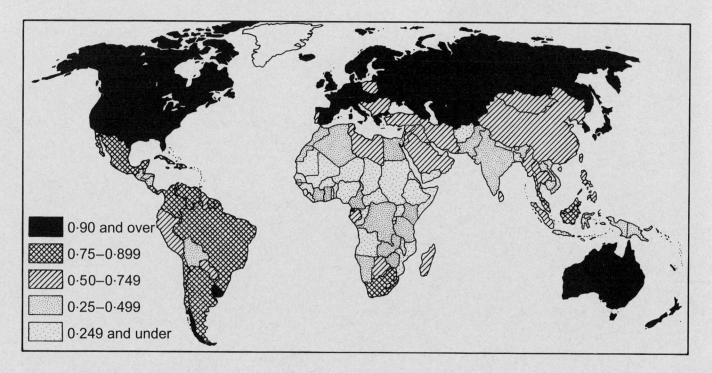

The Human Development Index measures development by combining three individual measures.  These measures are:

- adult literacy rate;

- life expectancy;

- real Gross Domestic Product (ie what an income will actually buy in a country).

*[END OF QUESTION PAPER]*

[BLANK PAGE]

[BLANK PAGE]

# Acknowledgements

Permission has been sought from all relevant copyright holders and Bright Red Publishing is grateful for the use of the following:

The U.S. Census Bureau for statistics from their website (2008 Paper 1 page 7);

A graph of global air temperatures 1855–2005 © Climatic Research Unit (2008 Paper 1 page 6);

An extract from www.bangor.ac.uk © Dr Ben Fisher (2006 Paper 2 page 2);

A photograph of a John Deere combine harvester © John Deere (2006 Paper 1 page 5);

A photograph reproduced by permission of Buchanan Galleries (2009 Paper 1 page 10);

A photograph reproduced by permission of Braehead shopping centre. (2009 Paper 1 page 10);

Ordnance Survey © Crown Copyright. All rights reserved. Licence number 100049324.